# ENERGY SAVINGS BY WASTES RECYCLING

*A report prepared for the Commission of the European Communities, Directorate-General Environment, Consumer Protection and Nuclear Safety, by Environmental Resources Limited, London, UK*

# ENERGY SAVINGS
# BY WASTES RECYCLING

*Edited by*

## RICHARD PORTER and TIM ROBERTS

*Environmental Resources Limited, London, UK*

Taylor & Francis
Taylor & Francis Group

LONDON AND NEW YORK

Taylor & Francis,
2 Park Square, Milton Park, Abingdon, Oxon, OX14 4RN

Transferred to Digital Printing 2005

**British Library Cataloguing in Publication Data**

Energy savings by wastes recycling: a report—
(EUR; 9222EN)
1. Recycling (Waste, etc.)
I. Porter, R.    II. Roberts, T.
III. Commission of the European Communities,
Consumer Protection and Nuclear Safety
IV. Environmental Resources Ltd.
604.6      TD794.5

ISBN 0-85334-353-5

WITH 70 TABLES AND 13 ILLUSTRATIONS

© ECSC, EEC, EAEC, BRUSSELS AND LUXEMBOURG, 1985

Publication arrangements by Commission of the European Communities, Directorate-
General Information Market and Innovation, Luxembourg

EUR 9222 EN

LEGAL NOTICE

Printed and bound by Antony Rowe Ltd, Eastbourne

# Executive Summary

This report contains the findings of a study which examines the potential energy savings available to the EEC through the further recovery of selected waste materials. Energy savings through materials recovery can arise:

—through substituting secondary materials for primary materials in the production of finished products;
—using recovered materials as a direct energy source.

For the materials considered in detail—aluminium, plastics, rubber, glass, waste paper and wood—the main opportunities for further waste recovery were identified as residing with municipal and other post-consumer wastes. Industrial wastes arising in material-forming and fabricating activities are already recycled to a high degree. However, in the case of wood there is a large potential for the further recovery of industrial wastes. There are considerable commercial, technical, locational, market and attitudinal constraints which will need to be overcome, however, if significant additional wood recovery is to be achieved.

Subject to this, additional energy savings of 564 PJ per annum (12·8 mtoe) are technically realisable. This is the saving accruing to the EEC alone and excludes the proportion of the total energy savings which would arise outside the EEC in raw materials extraction and processing. The figure is equivalent to 6 % of Community industrial energy consumption.

The potential energy saving has been estimated assuming the optimum combination of material recycling and direct thermal recovery. On this basis the energy-saving potential within the EEC for each material is (PJ per annum):

|  | Material recovery | Thermal conversion |
|---|---|---|
| Aluminium | 62 | — |
| Plastics | — | 135 |
| Waste paper | — | 17 |
| Glass | 23 | — |
| Rubber | — | 11 |
| Wood | — | 316 |
|  | 85 | 479 |

The major opportunities for energy recovery clearly reside with thermal conversion (85%), particularly of wood and plastics wastes. In the case of material recovery, aluminium provides the major energy-saving opportunity.

# Acknowledgements

The authors gratefully acknowledge the following for permission to reproduce: Fig. 5.5(c), Assovetro and the Italian National Research Centre; Fig. D1(a), from *Conservation and Recycling*, 1980, **3**, 260–9, Pergamon Press Ltd, England; Fig. D3.2(a), Foster Wheeler Power Products Ltd, England.

## CONTENTS

# 1. INTRODUCTION

## 1.1 Study Objectives

The overall aim of this study was to assess the energy conservation potential obtainable through the reuse of waste materials, and to identify areas for priority action.

The principal objectives of the study were:

o        to compare energy consumption per unit of production for industries using waste product/secondary materials as a raw material with energy consumption using natural raw materials;

o        to identify the extent of savings currently being realised and the scope for additional recovery, indicating those industries with most economic potential;

o        to determine the value and quantity of the energy savings in terms of crude oil equivalent imports potentially realisable through the recovery of secondary materials from industrial and municipal waste;

o        to identify the principal institutional, operational and technical constraints to further recovery;

o        to suggest what Community action would be necessary to overcome constraints and promote additional recovery, thereby leading to improved energy savings.

The original objective of the study was to assess the scope for additional energy savings resulting from further recycling which was **financially attractive, compared to alternative disposal options.**

However, the scope for additional secondary material recovery under strict financial criteria is very limited. On the basis of existing commercial technology and market conditions, the current level of material recycling in most industrial sectors can be considered close to the maximum which is financially viable. The secondary markets for most materials are well established and organised and can be expected, in broad terms, to maximise supply at a given price. In other words, the prices of primary materials would have to rise substantially on a permanent basis, relative to secondary materials, for significant additional recovery to take place.

We have therefore broadened our approach to consider what additional energy savings could be achieved through further technically feasible recycling of secondary materials. While most of the additional materials recovery potential identified could not be justified against commercial criteria, wider economic and social benefits would accrue. Such benefits include reduced energy consumption for a given level of economic output in the Community, lower energy and other raw material imports, raw material conservation, environmental benefits of reduced levels of waste disposal and, possibly, increased employment.

## 1.2 Scope of Study

Energy savings through materials recovery and reuse can arise in essentially two ways:

o       through substituting secondary materials for primary raw materials in the production of finished products;

o       using recovered materials as a direct source of fuel.

To determine the potential for such energy savings the study has identified:

o       the total quantities of municipal wastes in the Member States, their principal components and calorific value;

o       the total quantities of industrial and other wastes in the Member States by type and source;

o       industrial processes able to utilise secondary materials and direct energy recovery processes.

The materials included for initial evaluation were:

o       ferrous metals
o       non-ferrous metals
o       waste oils
o       solvents
o       plastics
o       rubber
o       glass
o       waste paper
o       wood

It was agreed following the Interim Report to concentrate the detailed analysis on aluminium, plastics, waste paper, glass, rubber and wood. These products were considered to have the greatest additional potential for energy savings, either because of the quantities still available to be recovered or because of the high unit energy savings achievable.

The energy savings which have been estimated are the incremental energy savings; that is, they are the additional energy savings which are still potentially available within the Community from further materials recovery. To put these savings in context we have also estimated the level of energy savings currently being obtained.

The energy savings have been estimated for the maximum practical
level of wastes still available for recovery, namely, the
quantity available after allowing for geographical dispersion and
handling and processing losses.  The extent to which this
potential can be achieved will depend not only on market factors,
such as the price of primary raw materials in relation to the
cost of recovering secondary materials, but also on how
effectively financial, institutional and social constraints can
be overcome.

### 1.3    Approach

In carrying out this study we have:

o          reviewed published data and studies concerning energy
           savings from secondary material recovery and the thermal
           processing of wastes;

o          held discussions with industry, government and waste
           disposal authorities, trade associations and individual
           experts;

A full list of the organisations and experts contacted made
during the course of the study are contained in Appendix F.

### 1.4    Report Layout

The report is presented in eight sections with Appendices:

Section 2 contains the principal findings and conclusions of the
study relating to energy savings.  The main factors affecting
further recovery are summarised and the areas for priority action
identified if further energy savings through secondary material
recovery are to be achieved.

Sections 3-8 contain the supporting evidence to the main
conclusions of the study as follows:

o          Section 3 summarises the quantities of waste arisings
           and the extent of current recovery activity;

o          Section 4 examines the different methods of recovering
           wastes, both as a material and as a fuel, and identifies
           the constraints to further recovery;

o          Section 5 describes the energy consumption of primary
           and secondary production methods and the potential unit
           energy savings available in different industries;

o          Section 6 determines the unit energy values available
           from the thermal conversion of wastes;

o          Section 7 contains the calculation of the overall energy
           savings available;

o        Section 8 reviews the options for government action to
         promote further energy savings.


1.5      **Glossary**

To assist the reader in understanding the study's findings, we
provide a glossary of terms used in the report.

**Waste descriptions**

**Waste:**  Any material which is rejected by the holder as no longer
having any value.

**Household wastes** are wastes arising at individual dwellings and
normally collected by local authorities.

**Commercial or trade wastes** arise at shops, offices, restaurants
and institutions such as schools, hospitals, municipal buildings
etc.

**Municipal waste** comprises household and commercial/trade waste.

**Industrial waste** arises at industrial plant.  It includes **prompt**
waste which arises during manufacture of the material and **process**
waste which arises during later stages of product fabrication.

**Post-consumer waste** includes products such as cars, appliances
and machinery which are not collected regularly by municipal
disposal authorities.

**Total waste arising:**  All waste materials generated at source,
whether subsequently recovered or not.

**Dissipative losses:**  These losses occur in use of the commodity
and render recovery impracticable.

**Resource recovery** is the process of collection, transfer,
sorting, separation, decontamination and/or upgrading etc., of
materials in waste for the purpose of recycling, reuse, by-
product generation or use for energy recovery or compost.

**Recycling** is the recovery and subsequent use of a secondary
material for the manufacture and/or fabrication of a material or
product similar to that from which the secondary material
originated.

**Reuse** is recovery of a secondary material and its subsequent use
in its original form and for its original use (e.g. as in returnable glass bottles).

**'New'** scrap comprises **prompt and process wastes** arising during
manufacture and processing of the material in fabrication of
articles.

**'Old'** scrap comprises wastes arising from **municipal and post-
consumer** sources but may include redundant plant arising in
**industrial waste**.

**Thermal conversion** includes all processes utilised to process combustible materials with the liberation of free energy.

**By-product generation** is the recovery from waste of a secondary material for fabrication of a material or use as a product different from that from which the secondary material originated (e.g. use of waste rubber in road surfaces).

1.6      **Units Used**

**The Tonne** - one tonne is a mass of 1,000 kg equivalent to 0.9842 long ton or 1.102 short tons.

**The Joule** (J) is the amount of heat required to warm 0.239 grammes of water through 1°C.

**The million tonne oil equivalent (mtoe)** is the thermal equivalent of one million tonnes of crude oil, which is equivalent to $41.9 \times 10^{15}$ J.

**Prefixes for SI units**

The following prefixes have been utilised in the report.

| Multiple | Prefix | Symbol |
|----------|--------|--------|
| $10^3$ | kilo | K |
| $10^6$ | mega | M |
| $10^9$ | giga | G |
| $10^{12}$ | tera | T |
| $10^{15}$ | peta | P |

i.e. $1 \times 10^{15}$ J  $=$  1 PJ

1.7      **Acknowledgements**

We would like to express our thanks to the many people in public and private organisations in Europe who lent us their time, data and comments during the course of the study. Their assistance is very much appreciated.

2.      CONCLUSIONS AND RECOMMENDATIONS

2.1     **Materials Recovery for Recycling and Thermal Conversion**

2.1.1   **Existing levels of material recovery**

The recovery of wastes is a well established practice throughout Europe. This is particularly true of wastes which arise in material forming and fabricating activities within industry (so-called prompt and process scrap) where recovery is already at a very high level. For instance, in the case of aluminium 99% of prompt and process wastes are recovered; in the case of plastics up to 88% of these wastes are recovered.

For the materials under detailed review only in the case of plastics (in small quantities) and wood were further industrial waste recovery opportunities identified.

The main opportunities for **further recovery therefore lie with municipal and other post-consumer wastes.**

2.1.2   **Methods of recycling municipal refuse**

The waste materials arising in municipal refuse can be recycled as secondary materials or, in the case of combustible materials, used as an energy source. In the case of recycling, the secondary materials may be separated:

o       either **at source** by the consumer; or

o       **mechanically,** in a municipal plant, after collection.

**Source separation:**  Such schemes are widely established throughout Europe for the recovery of a variety of materials - glass, cans, newspapers, office waste paper and plastic bottles. The materials are either separated at source for collection by a third party or separated at source and taken by the consumer to a collection point.

Collection costs are considerably less if undertaken by voluntary groups instead of by local authority collection crews. However, the transporting of small quantities of materials by many individuals (such as in bottle bank type schemes) may well be less energy efficient if special journeys have to be made.

A major consideration in the further development of source separation schemes is the extent to which householder collaboration can be increased. Consumer motiviation has traditionally relied on goodwill based on an appreciation of the environmental and conservation benefits which accrue. Whether the motivation is sufficient to raise further the level of source separation/ returning of secondary materials and whether a wider spectrum of social groups with a lower inclination to participate can be suitably motiviated, is uncertain.

Financial incentives are being offered in certain instances to encourage participation but there is a narrow, and in some circumstances non-existent, margin available for the recycling agency offering sufficient inducement and still retaining financial viability.

**Mechanical separation:** A wide variety of recovery methods for material recycling have been developed in recent years to reclaim components of mixed wastes by means of automatic sorting processes. Many of these have not been developed beyond the pilot and experimental stages and there is very limited experience of large-scale 'commercial' plant operations. From the experience to date of the three full-scale plants (one each in France, Germany and Italy) using various separation technologies, the technical and economic efficiency of these processes remains uncertain. While ferrous metal extraction poses no problems, the separation of aluminium, glass and plastics is far more problematical. The separated materials are generally contaminated with other wastes and manual separation is often additionally required.

Even if the technical problems facing these plants can be overcome they are only likely to be viable in urban areas where a sufficient volume of wastes can be generated to provide an economic base load.

## 2.1.3    Methods of direct energy recovery

Energy has been recovered for many years from domestic refuse through direct incineration methods. More recent approaches to combustion energy recovery of municipal waste have been to convert raw refuse into more conveniently handled fuels in solid form (refuse-derived fuels).

Following unsatisfactory experience with the first generation plants, there is also renewed interest in the pyrolysis and gasification of wastes which provide gases, oil and char. Landfill gas recovery technology is also well established; this, however, was not included for consideration in this study.

**Incineration:** Around 24% of municipal waste is currently incinerated in Europe - 16% with energy recovery. There are two main options for on-site energy recovery associated with incineration:

o        for heat, by producing steam, and distribution to consumers as low grade heat for district heating (or for industrial use);

o        for electricity, with a conventional boiler turbine for sale into the national grid.

Energy recovery from municipal waste is a capital intensive process; its feasibility depends critically upon the price received for the energy (usually electricity) produced.

**Refuse-derived fuel:** There are several full-scale RDF manufacturing plants (mainly producing pelletised fractions of municipal waste) in Europe. As their development has generally been subsidised, commercial viability (i.e. lowest net cost disposal method) is not yet proven although the outlook is encouraging. Currently around 2.84 Mt of municipal wastes is being converted to RDF, representing 3% of total municipal waste arisings.

In market terms, RDF competes principally with coal. It may be noted that there is an energy penalty the more that RDF is refined to produce a higher calorific value fuel.

**Pyrolysis:** The pyrolysis/gasification of wastes is still a development technology. In addition to laboratory scale systems and demonstration pilot plants operating there is only one full-scale plant currently operating in the Community - at Creteil in France. This process is of particular relevance to direct energy recovery from tyres and plastics.

Pyrolysis approaches municipal bulk incineration in terms of efficiency. After providing its own fuel needs it can provide steam for either plant/district heating or electricity generation.

### 2.1.4    Constraints to further recovery

**Materials recycling:** If significant further recycling and energy savings through reuse of secondary materials is to be achieved various technical/economic and other constraints will need to be overcome. Examples of such constraints are as follows:

o    industry and consumer attitudes to products incorporating recycled materials;

o    difficulty of maintaining a continuous supply of waste materials of required quality;

o    institutional and locational constraints;

o    problem of increasing consumer participation in resource recovery schemes;

o    the technical and economic constraints associated with recovery of more highly contaminated and mixed wastes;

o    vested industrial interests which prefer traditional primary raw material sources to secondary materials, (e.g. vertical integration);

o    fluctuating price movements in secondary material markets.

The development of new markets for secondary materials may also be a necessary critical development, especially for those materials such as plastic and wood wastes which, for technical reasons, are often not able to substitute directly for equivalent virgin materials but do have the potential to substitute for certain lower grade primary materials.

**Thermal recovery:** Institutional and locational factors are particularly significant in relation to the recovery of the energy content of secondary materials through direct combustion/conversion processes. In this situation, constraints can arise in one or more of the following ways:

o       borrowing/financial constraints on local authorities;

o       area of responsibilities/role or tradition in community/ municipal enterprises;

o       location and size of energy consumers;

o       difficulty of maintaining a continuous supply of combustible material of consistent calorific value;

o       price obtainable for selling heat/electricity surpluses to local utility/CHP network.

## 2.2    Energy Savings through Additional Recovery

### 2.2.1    Unit energy savings through recycling

Energy savings result from waste recycling if the energy used in collecting, separating and treating reclaimed wastes, and subsequent processing, is less than the energy used in originating and processing primary materials and disposing of wastes (see Figure 2.2(a)).

There are considerable variations in material production processes both within and between countries. Also the age of plant and purity and form of the primary and secondary materials affects the energy consumption of manufacturing processes. In Figure 2.2(b), we compare typical energy consumption per unit of production using primary and secondary materials, indicating the principal energy consuming components.

Plastic has not been included since it is not a substitute material in the same sense. It is extremely difficult to substitute recovered used plastics for virgin polymers. In so far as waste plastics can be reused it is generally in low grade products as substitutes for other materials. Nonetheless, recovery and reuse of plastics for any use can offer energy savings and the plastic is still available for eventual use as a fuel source.

Similarly secondary wood products are not strictly substitutable for primary wood products as both have differing uses and characteristics. Furthermore, the energy required to produce secondary wood materials usually exceeds the primary energy consumed in the production of the finished primary material.

11

FIGURE 2.4(a):   STAGES WHERE ENERGY CONSUMED - PRIMARY AND SECONDARY
MATERIAL CYCLE

**Primary Material Route**

**Secondary Material Route**

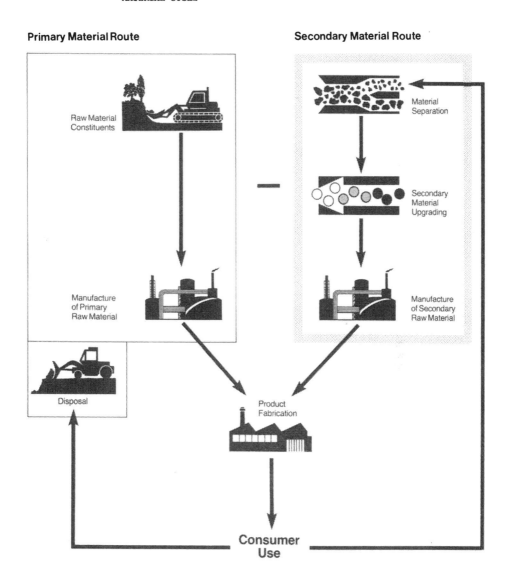

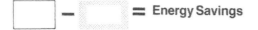 = Energy Savings

12

**FIGURE 2.2(b): COMPARISON OF UNIT ENERGIES FOR PRIMARY AND SECONDARY PROCESSING OF FOUR MATERIALS**

**ALUMINIUM**

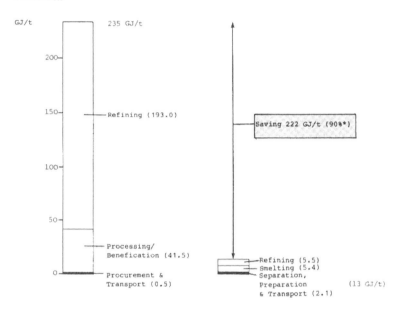

Primary Processing

Secondary Processing

*actual 94% saving reduced by consumer energy and primary production efficiency increases.

**WASTE PAPER**

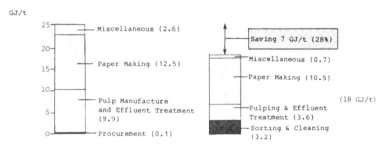

Primary Processing

Secondary Processing

**FIGURE 2.2(b): continued**

**GLASS**

GJ/t

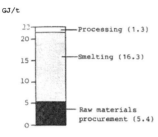
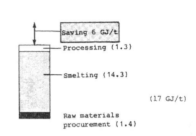

**Primary Processing**

**Secondary Processing**
(excludes consumer energy)

**RUBBER**

GJ/t

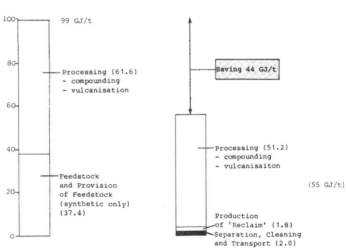

**Primary Processing**          **Secondary Processing**

From the comparative energy content analysis above, the following conclusions may be drawn:

**Aluminium**: By far the greatest unit energy savings are achieved by aluminium conventionally in the form of electricity at the smelting stage. In theory, the primary equivalent of reduced electricity consumption will depend on whether hydro- or thermal-power generation is involved. In practice, it is reasonable to assume that electricity savings at the margin by substituting secondary aluminium in the EEC would globally reduce unit energy input by about 90%.

**Paper**: The energy saving through recycling and use of waste paper varies considerably by quality of waste paper and type of end product. To the extent that further recovery of waste paper will be from more contaminated sources, requiring energy intensive preprocessing (such as deinking), the unit energy savings potential will be restricted. Also to utilise all the wastes, new markets will need to be found where the energy substitution effect is uncertain.

**Glass**: A maximum 6 GJ/tonne energy saving is available in producing glass from cullet rather than primary raw materials. This assumes a 100% cullet utilisation in glass production compared to the current level of around 20%. While 100% utilisation of mixed cullet may be feasible in the production of certain types of coloured glass, recent experience suggests 60% as the maximum utilisation level of mixed cullet from domestic waste.

More impressive are the energy savings associated with returnable bottles. The extent of the saving clearly depends on the trippage rate (the number of times the bottle is reused). Assuming a trippage rate of 10, energy savings of 45% can be expected.

**Rubber**: The energy savings in the recycling of rubber arise in the production and moulding of 'reclaim and crumb'[1] compared to rubber production from primary materials. In addition to process energy savings, recovered rubber replaces the oil feedstock content of synthetic rubbers, which amounts to about 40% of their total energy content. The 45% energy saving shown in Figure 2.2(b) is the weighted average of using secondary instead of primary raw materials in rubber manufacture, taking account of the different grades of rubber, principally the proportion of synthetic versus natural.

In respect of the retreading of tyres, a major rubber recovery activity, the unit energy savings relative to new tyre manufacture are around 66%.

---

[1] 'Crumb' is waste rubber after it has been physically broken down into granular material. 'Reclaim' is secondary rubber which has also been chemically treated before any revulcanisation.

In the estimates of energy consumed per unit of product manufactured, no account has been taken of energy consumed in **the collection of waste/secondary materials.** This type of energy consumption is common to primary and secondary material processing options. If additional transport energy is consumed in collecting waste separated at the point of arising above that which would be involved in collection of non-separated municipal or consumer waste, then this should be taken into account. It can be argued that this incremental energy cost is small or negligible and will depend on which one of the many different types of recovery collection systems is used. The energy usage could be significant however where bottle bank-type schemes are operated and where the consumer makes a special journey to return the bottles.

Reliable statistics on consumer habits in this respect do not exist but if, in the case of glass bottles for example, a special journey to return the source of cullet is made, the energy consumed is around 10-12 GJ/tonne of glass, i.e. twice the maximum unit energy saving obtained from substituting cullet for primary raw materials in glass manufacture.

2.2.2    **Comparison of unit energy savings from recycling and direct energy recovery processes**

For combustible wastes the option exists to save energy either through recycling or through direct energy recovery processes such as incineration with heat recovery/electricity generation or pyrolysis. It is of interest to know for paper and rubber which of the two alternative recovery routes yields the greater energy savings benefit. This is illustrated in Figure 2.2(c). It may be seen that after account is taken of the fact that:

o        a proportion of the potential energy savings in recycling waste materials is likely to lie outside the EEC (see Section 2.3) - 30% in the case of both waste paper and rubber;

o        the energy conversion processes utilising waste materials are roughly only 70-85% as thermally efficient as processes burning the fossil fuels being displaced (saved);

then, in the case of waste paper, the direct thermal conversion route offers the greater unit energy saving [1]; for rubber material recycling yields the greater unit energy saving.

_____

[1]  In theory, the energy savings from recycling might be thought of as being obtainable more than on a 'once through' basis. In practice, this is unlikely.

2.3    **Total Community Energy Savings Potential from Further Secondary
       Material Recovery**

2.3.1  **Mitigating factors**

The theoretical energy savings potential in the Community from
further materials recovery is obtained by multiplying the
technically recoverable and reusable quantities of
wastes/secondary materials by the unit energy savings available.

The practical level of potential energy savings available however
would be something less than this because:

  i.     it is not even remotely economic to consider collecting
         and aggregating the 20% or so municipal/post-consumer
         wastes arising in isolated rural areas;

  ii.    there are handling and process losses associated with
         the pretreatment and use of secondary materials as
         industrial feedstocks, estimated to be 15-35% depending
         on the material involved;  these losses are not incurred
         with direct thermal conversion processes;

  iii.   a proportion of energy savings in the raw material
         extraction and manufacturing process phases of certain
         products occur outside the EEC.

With regard to the first two factors, appropriate downward
adjustments have been made to determine the net waste arisings
still available for recovery.

In respect of energy savings occurring outside the EEC, unit
savings adjustments have been made based on the current
proportions of total finished materials consumed in the EEC for
which the primary raw materials are extracted and, to some
extent, processed outside the Community.  The proportion of unit
energy savings which are estimated to occur outside the Community
are as follows:

              Aluminium    50%
              Waste Paper  30%
              Glass         3%
              Rubber       30%

The net unit energy savings after making this adjustment are
shown in Table 2.2(c).

2.3.2  **Estimated total energy savings potential**

Currently, it is estimated that total energy savings realised
through existing levels of recycling and direct waste/energy
recovery processes is 1,356 Petajoules (PJ per annum), or about
3.2 million tonnes oil equivalent (mtoe).

FIGURE 2.2(c): COMPARISON OF UNIT ENERGY SAVINGS BY MATERIAL RECYCLING
AND THERMAL CONVERSION BY BULK INCINERATION

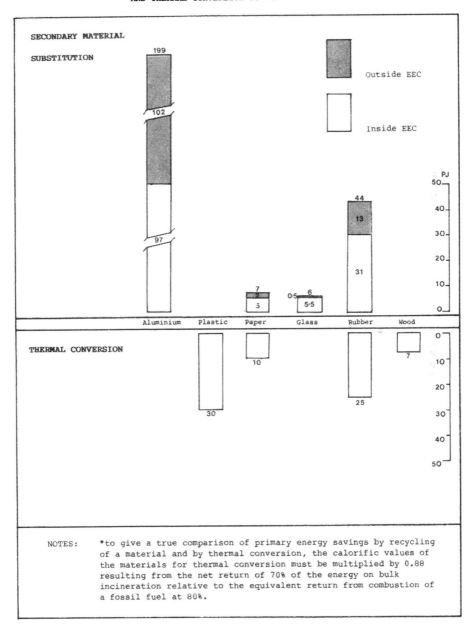

NOTES:     *to give a true comparison of primary energy savings by recycling
of a material and by thermal conversion, the calorific values of
the materials for thermal conversion must be multiplied by 0.88
resulting from the net return of 70% of the energy on bulk
incineration relative to the equivalent return from combustion of
a fossil fuel at 80%.

In Figure 2.3(a) we show the total gross energy savings potential which would be realised through the optimum combination of recycling and direct thermal recovery. It should be stressed that these represent the <u>maximum</u> obtainable energy savings potential under currently known technology after taking account of the mitigating factors identified in Section 2.3.1. They would mostly not be seen as financially attractive under current market conditions, and in many cases would require other industrial, consumer and institutional constraints to be overcome.

Even with external subsidies, it is unlikely that the full potential of 564 PJ/a (12.8 mtoe) could be realised. Nevertheless, it shows the order of magnitude of energy savings potential. While this figure in the Community represents only some 6% of industrial energy consumption, it may be seen that it is a similar order to certain other energy savings potential realisable through various conservation measures.

For example, studies carried out for the European Commission have revealed that:

o        in the provision of space heating in domestic, commercial and industrial buildings, the introduction of heat pumps utilising ambient energy in the environment would yield annual fuel savings of 25-35 mtoe/year [1];

o        in the transport sector, the introduction of speed limits (90 km/hour) could produce 10-15% fuel savings worth 14-21 mtoe in 1980 Community transport fuel consumption. Similar reductions are thought possible through reducing vehicle weight, size and power, optimising engine performance, gear ratios, vehicle aerodynamics etc. [2]

It may be seen that a large share of the estimated energy savings is accounted for by the potential for recovering the thermal content of wood wastes and, to a lesser extent, of plastic wastes. The latter reflects the considerable quantities of unrecovered plastics in municipal wastes. The large quantity of industrial wood wastes for which energy recovery potential exists is to be found principally in forest harvesting industries, where considerable economic constraints would limit the realisation of the full potential.

---

[1]  Evaluation of the Community Demonstration Programme in the Energy Sector. CEC DGE XVII/132/82-EN.

[2]  Energy Savings in the EEC. Jean Leclercq et al. Agence European d'Informations. 449/205, 1981.

**FIGURE 2.3(a): GROSS ENERGY SAVINGS POTENTIAL BY APPROPRIATE RECYCLING OR THERMAL CONVERSION PROCESSES**

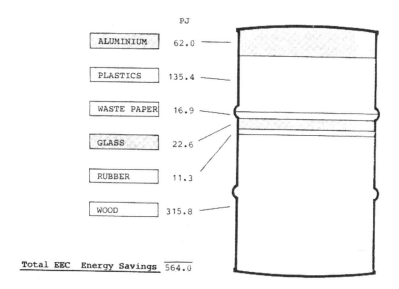

PJ

| | |
|---|---|
| ALUMINIUM | 62.0 |
| PLASTICS | 135.4 |
| WASTE PAPER | 16.9 |
| GLASS | 22.6 |
| RUBBER | 11.3 |
| WOOD | 315.8 |

Total EEC  Energy Savings  564.0

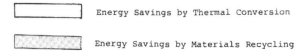

☐  Energy Savings by Thermal Conversion

▨  Energy Savings by Materials Recycling

2.4     **Action Necessary to Realise Additional Energy Savings**

2.4.1   In this section we identify the areas for action if additional significant energy savings obtainable are to be realised through secondary materials recovery. Suggestions are also made as to those areas where the Commission may be able to take initiatives.

It should be emphasised that we have specifically concentrated on those sectors for which further secondary material recovery and reuse would have a significant impact on energy use. There are, of course, other wider strategic, economic and environmental benefits to be gained from materials recycling which may also further support or justify wider action to encourage materials recycling and development of waste/energy recovery processes.

2.4.2   **Materials recycling**

**Aluminium:** From an energy savings viewpoint, aluminium stands out as the material for which further recycling would achieve major energy benefits. Further aluminium recovery depends on progress being made in separation techniques and improved collection systems for dispersed scrap in post-consumer wastes (particularly durable goods) and for demolition wastes. The current levels of aluminium in municipal waste are still small, although likely to grow with increasing adoption of the all-aluminium can. Source separation schemes would seem to offer the best opportunity for increased recycling. It is suggested, therefore, that:

   i.     governments should support studies to identify more clearly where and in what form aluminium post-consumer waste principally arises; the Commission could lend its own support to these investigations and encourage the dissemination of the findings;

   ii.    clearer assessments are needed of the feasibility of separate waste collection schemes of household and other source aluminium wastes; incentives may well have to be offered by central government to municipal and other authorities for the development of such schemes;

   iii.   the Commission should consider providing financial support for research and development, including demonstration projects, in technologies for separating and upgrading aluminium scrap from other materials.

**Glass:** Increased recovery (as cullet) is dependent upon further public participation in bottle bank schemes, and the willingness of local glass plants to receive cullet. However, it should be recognised that the identified potential energy savings to be achieved by further glass recycling are partly dependent upon the collection/return of bottles not involving special trips to the bottle bank on the part of the consumer.

i.    It is recommended, therefore, that before further action
      is taken to encourage additional recycling of glass as
      cullet, social surveys be undertaken in Member States to
      assess the existing consumer behaviour patterns with
      respect to transport/return of bottles, and likely
      future response to and use of private transport in any
      expansion of bottle bank schemes.

ii.   If the consumer surveys indicate little additional
      collection/delivery transport energy is likely to be
      involved in further recovery of glass, encouragement
      should be given by central and regional/municipal
      governments, also involving glass manufacturers, to the
      introduction of bottle bank schemes.

Whatever the outcome of such investigations, there would also
seem to be strong grounds for:

iii.  encouragement of reuse of glass containers - in this
      respect any steps to facilitate reuse, including the
      standardisation of such containers as is currently under
      consideration by the Commission, should be supported.

**On the grounds of energy saving alone,** other action on the part
of the Commission or Member State Governments to encourage
further recycling of the secondary materials considered in detail
here, would not seem justified.

**2.4.3**    **Direct energy recovery/thermal conversion of secondary materials**

**Thermal conversion of waste paper, plastics and rubber in
municipal and other post-consumer wastes:**  There are significant
further energy savings potentially achievable through the
realisation of combustible energy content of municipal and other
post-consumer wastes:

|             | Municipal Wastes | Other Post-Consumer Wastes |
|-------------|------------------|----------------------------|
| Plastics    | 29               | 103                        |
| Waste Paper | 15               | -                          |
| Rubber      | 6                | 5                          |
| Wood        | 1                | 21                         |
|             | 51 PJ/a          | 129 PJ/a                   |

For plastics in particular direct thermal conversion processes,
rather than materials recycling, offers the most energy efficient
means of realising this potential.

These findings lend strong underlying energy savings rationale to
the recommendations contained in the recent reports to the
Commission on Waste/Energy Conversion Technologies, in particular
with regard to the:

o     further development of refuse derived fuel systems;

o      incineration with energy recovery and,

o      pyrolysis of municipal wastes (in suitable circumstances).

Attention should be given by central governments to ensure that institutional barriers, in particular unfair or inflexible electricity purchasing policies by utilities from small producers, do not hinder the development of municipal waste energy recovery schemes.

**Industrial waste incineration/pyrolysis**: Very considerable energy savings potential lies in the energy content of unrecovered wood wastes (294 PJ), and to a much lesser extent, mixed plastic wastes (3.4 PJ). For the plastics and contaminated wood wastes on-site small-scale incineration (e.g. starved air technology systems) with energy recovery offers the most suitable and economic option. The principal constraints to further development of this technology can include:

o      availability of capital resources on the part of industrial energy consumers;

o      in the case of pyrolysis, resistance on the part of industry to use unfamiliar waste combustion technologies;

o      fluctuating supply/availability of these wastes material, increasing/variable load on the energy plant;

o      in the case of wastes from forest product industries, the relatively limited number of near-by industrial/ commercial energy users capable of utilising the wood wastes, although considerable potential exists within the forest product industry itself, if surplus electricity generated could be sold to the public supply system.

The first two constraints could be partly overcome through the making capital grant/financial incentive schemes available to industry.

The majority of wood wastes, which are uncontaminated could be satisfactorily combusted in suitable solid fuel boilers. Conversion of existing solid fuel boilers would cost relatively little; conversion of oil fired boilers, if not physically impracticable, would be of high cost.

Pyrolysis, and also incineration inside cement kilns, offer the most suitable technical means for realising energy content in waste rubber. Economic development and, therefore justification to encourage such schemes by government, are likely to depend upon local circumstances. This could be encouraged by further support of demonstration projects/firing trials for application of such technologies.

3.    **WASTE QUANTITIES AVAILABLE FOR RECOVERY**

3.1    **Introduction**

In this section we summarise for each material our estimates of:

o        waste arisings;

o        quantities of waste currently recovered;

o        potential for further recovery.

The materials initially examined were aluminium, plastics, waste
paper, glass, rubber, ferrous and non-ferrous metals, waste oils,
solvents and wood.  Information on waste arisings and current
recovery levels of these materials are contained in Appendix A.
Of these materials aluminium, plastics, waste paper, glass,
rubber and wood have been selected for detailed analysis.  These
products were considered to have the greatest additional
potential for energy savings.

In Table 3.1(a) the estimated waste arisings and quantities
currently recovered in the EEC from municipal and post-consumer
wastes are given for selected waste categories.  The physical
availability estimate given in the table represents the maximum
quantity of wastes still theoretically available for further
recovery.

| Table 3.1(a) | | | |
|---|---|---|---|
| ESTIMATED WASTE ARISINGS AND CURRENT RECOVERY IN THE EEC 1981 - SELECTED WASTE CATEGORIES (All data Mt) | | | |
| | Estimated Arisings(1) | Currently Recovered (2) | Physically Available for Further Recovery |
| Aluminium | 1.479 | 0.385 | 1.094 |
| Plastics | 11.590 | 1.710 | 9.880 |
| Waste Paper | 28.065 | 12.000 | 16.065 |
| Glass | 8.910 | 1.810 | 7.100 |
| Rubber | 1.650 | 0.500 | 1.150 |
| Wood | 19.500 | 0.500 | 19.000 |
| (1)    Data sources given in Appendix A. (2)    Excludes new scrap (prompt and process wastes) Excludes wastes incinerated with heat recovery. | | | |

The supporting data and the methododology adopted for making
these estimates are contained in Appendix A. This Appendix also
provides details of arisings and recovery by waste category and
country.

In Appendix B a fuller qualitative review of existing recovery
activities and the main opportunities for further recovery is
given. Below we summarise the salient features of recovery
activity for the waste categories under review.


3.2       **Aluminium**

3.2.1     **Waste arisings and current recovery levels**

The total arisings of aluminium in the EEC in municipal and post-
consumer wastes are estimated as 1.5 million tonnes, of which
0.385 million tonnes are currently recovered.

The main end uses and current recovery activity for aluminium are
summarised in Table 3.2(a).

The recovery of waste aluminium is well established; around 28%
of the aluminium processed in Europe is recycled metal. There is
every incentive to maximise aluminium recovery as only 12% of
consumption is available from local mining production.

**New scrap:** There is a very high level of recovery of prompt and
process wastes. This 'new' scrap is clean scrap arising in-plant
either from initial forming or from final fabricating activities.
It takes the form of trimmings, punchings, borings, imperfect
products, drosses, sweepings, etc. Because the scrap is clean
and homogeneous and arises in large amounts at specific
locations, the economics of recovery are highly favourable. Such
scrap finds a ready market for resale either to the primary or
secondary metal industries and recovery is thought to approach
100%. More than 60% of the industry's scrap waste is 'new'
scrap. The opportunities for further recovery of these wastes
are very limited.

**Old scrap:** The recovery of old aluminium scrap from household
and post-consumer wastes is less developed. Post-consumer wastes
arise as part of transport vehicles, consumer durables, office
equipment, industrial plant and miscellaneous wrought and cast
aluminium products. The recovery of aluminium from domestic
wastes (mainly packaging material - cans and foil) is at a much
lower level, partly reflecting the small proportion which
aluminium has traditionally represented of household waste -
often not exceeding 1% by weight. With the increased use of
aluminium for beverage cans however more attention is being given
to recovery possibilities from this source.

| Table 3.2(a) | | |
|---|---|---|
| ALUMINIUM - MAIN USES AND RECOVERY ACTIVITY | | |
| End Uses | Processed Forms | Current Recovery Activity |
| Transport<br><br>Mechanical<br>Engineering | **Pure aluminium** -<br>mainly foil and<br>principally in<br>packaging industry. | **In-plant prompt scrap<br>recovery:**High level of<br>recovery - approaching<br>100% |
| Electrical<br>Engineering<br><br>Construction<br><br>Chemicals | **Casting alloys** -<br>aluminium alloyed<br>with silicon, copper,<br>magnesium etc., for<br>casting purposes. | **Process plant scrap<br>recovery:** High level of<br>recovery - approaching<br>100% |
| Packaging<br><br>Consumer<br>Durables<br><br>Steel<br>Alloys<br><br>Destructive<br>Uses | **Wrought alloys** -<br>are rolled, drawn,<br>extruded or formed<br>by some fabricating<br>method other than<br>casting, to produce<br>sheet, wire, forgings,<br>extrusions or tubes.<br><br>**Other forms** - use of<br>aluminium in alloys<br>and in production of<br>aluminium chloride<br>and explosives. | **Recovery of old scrap:**<br>Post-consumer scrap<br>(consumer durables,<br>capital goods, cars);<br>Recovery of post-<br>consumer waste widely<br>practised.<br>Domestic and consumer<br>wastes (packaging):<br>Recovery of packaging<br>from domestic and<br>commercial wastes at very<br>low level. |

With very few exceptions, aluminium scrap arises in conjunction
with other materials, requiring separation before the aluminium
scrap can be used.  Its high value however means that mechanical
and other collection and sorting systems can often be used on a
commercial basis without economic constraints.  The highest scrap
recovery rates for aluminium products occur for concentrated cast
aluminium parts in automobiles, plant and in domestic and office
equipment, and for wrought aluminium products that arise in
isolation, such as litho sheets and beer barrels.

It is the recycling of obsolete aluminium scrap that offers the
major opportunity for the further conservation of material and
energy resources.

## 3.3    Plastics

### 3.3.1    Waste arisings and current recovery levels

The total arisings of plastics in the EEC in municipal and post-consumer wastes are estimated as 11.6 million tonnes, of which 1.7 million tonnes are currently recovered.  The main end uses and current recovery activity for plastics are summarised in Table 3.3(a).

| Table 3.3(a) | | |
|---|---|---|
| PLASTICS - MAIN USES AND RECOVERY ACTIVITY | | |
| End Uses | Processed Forms | Current Recovery Activity |
| Building<br><br>Packaging<br><br>Electrical Products<br><br>Automotive Products<br><br>Furniture | Monomers are chemically combined to produce polymer resin which may then be compounded with additives to change properties, e.g. make more rigid, flexible, heat-resistant, opaque, etc.<br><br>**Thermoplastic** resins provided to fabricators in form of lattices, pellets, powder or granules for shaping either by injection moulding, blow moulding, extrusion or thermo-forming.<br><br>**Thermosetting** resins provided in granular or liquid form and are usually processed by compression, transfer moulding, casting or calendering. | **In-plant prompt scrap recovery** by resin manu-facturers; high level of recovery.<br>**Process plant scrap recovery** by fabricators and converters - or independent reprocessors; high level of recovery.<br><br>**Recovery of old scrap** - post-consumer scrap (consumer durables, capital goods, cables); low level of recovery. |

**New scrap:**  It is standard practice amongst resin producers and most large fabricators to gather, recycle and rework as much scrap as possible in their processes.  Scrap plastic is produced at every stage of plastics manufacture.  Resin producers have polymer wastes resulting from emergency dumps of their reactors and auxilliary equipment.  Resin producers, fabricators and converters all generate wastes from off-grade products, spillage and equipment cleaning.  Fabricators and converters also produce waste from trimming, moulding and forming operations.

Certain of the smaller fabricators and converters do not find it economic to employ in-plant reclamation. In such cases, independent reprocessors collect, grind and blend these wastes for recycling as a secondary resin. Both internal and external recycling of prompt and process wastes (apart from highly mixed or contaminated material) is thus developed to a high degree.

**Old scrap:** In contrast, the recovery of plastic from post-consumer wastes is at a very low level. The major proportion of plastics waste in refuse derives from discarded packaging material (around 90%). Other non-packaging plastic products are retained in consumption for a comparatively much longer period of time (anything up to 40 years compared with less than a year for packaging plastics). In the last 10 years, plastics have become an increasingly important material for packaging purposes.

Thermoplastics are by far the most important plastics used in packaging and polyolefins are the most common of the thermoplastics. There is a considerable amount of research and development work being undertaken throughout Europe on the recovery of the plastics fraction of domestic waste but very little is recovered at present.

There are technical factors which hinder plastics recycling:

o        the production of thermosetting plastics requires the setting into permanent shape of the resin by the application of heat and pressure. These plastics are thus by their nature impossible to recycle as a material, though they do offer limited potential for by-product generation;

o        thermoplastics are frequently mixed with plastics based on different resins or with completely different materials during the course of fabrication and conversion. As different thermoplastic polymers are incompatible they must be separated before they can be used for re-processing. Further difficulties are caused by the mixing of plastics with other non-plastic materials, e.g. metal composites, e.g. in cars, paper, foil-backed packaging film and refrigerators.

Effective recovery technology declines rapidly as the virgin resin moves towards the end-use item. The more mixed, the dirtier, the wetter and the less consistent the scrap, the cruder and less consistent are the products that can be made from it and the greater the cost of converting it.

## 3.4        Paper

### 3.4.1        Waste arisings and current recovery levels

The total arisings of waste paper in the EEC in municipal and post-consumer wastes are estimated as 28.1 million tonnes of which 12.0 million tones are currently recovered. The main end uses and current recovery activity for waste paper are summarised in Table 3.4(a).

| Table 3.4(a) | | |
|---|---|---|
| WASTE PAPER AND BOARD - MAIN USES AND RECOVERY ACTIVITY | | |
| End Uses | Processed Forms | Current Recovery Activity |
| Newsprint<br><br>Other Printing & Writing Paper | **Mechanically** produced pulp is used mainly in products not designed for permanent use, e.g. newspapers and magazines. | **In-plant prompt waste recovery:** High level of recovery. |
| Packaging Paper and Board<br><br>Construction Paper and Board | **Chemical** pulp (e.g. Kraft pulp) is stronger than mechanical pulp and is used extensively in brown paper manufacture (e.g. carrier bags, corrugated board, etc.) | **Process plant recovery** by manufacturers, converters and printers: High level of recovery. |
| Household & Sanitary Paper | Writing paper can be made from all qualities of pulp. Tissues are made from chemical pulp. Board manufacture utilises a high proportion of waste paper, often in a laminate structure. | **Recovery of old waste** - newspapers, fibre board containers, mixed waste paper from domestic, trade and industrial sources: High level of recovery from industry and commerce. Variable level of recovery from domestic sources but generally not high. |

In the Community as a whole waste paper has grown in importance as a raw material to the paper industry in the last 10 years; it now provides 45% of the raw material used in paper production.

**New scrap:** As with the other materials being considered here, there is a high level of recovery of wastes arising during the manufacture and fabrication of paper. Production wastes generated at the mill are returned for re-pulping. Cuttings or sheets, plain or printed, arising at a fabricating plant are generally clean and homogeneous and are frequently accepted by the original manufacturer for recycling. The main potential for further recovery therefore once again rests with discarded end-products.

**Old scrap:** The main forms of waste paper and board occurring within domestic and trade/industrial wastes are:

o **domestic** - newspapers, fibreboard containers, mixed waste paper;

o **trade/industrial** - fibreboard containers and mixed waste from shops, stores, offices, warehouses and factories.

The bulk of household and sanitary papers, together with construction paper and board, are not reclaimable by virtue of their end-uses. The principal sources of paper in domestic and trade refuse are discarded packaging materials with newsprint, printing and writing paper predominating.

Waste paper is not a direct alternative to pulp in the manufacture of several high grades of paper, but in the production of a number of grades of packaging there is no economic substitute for waste.

## 3.5 Glass

### 3.5.1 Waste arisings and current recovery levels

The total arisings of glass in the EEC in municipal and post-consumer wastes are estimated as 9.0 million tonnes of which 1.8 million tonnes are currently recovered. The main end uses and current recovery activity for glass are summarised in Table 3.5(a).

Table 3.5(a)

| GLASS - MAIN USES AND RECOVERY ACTIVITY | | |
|---|---|---|
| End Uses | Processed Forms | Current Recovery Activity |
| Flat Glass (windows etc)  Containers (packaging & industrial)  Domestic (household & ornamental)  Glass fibre  Miscellaneous (safety, optical, laboratory, hygienic & pharmaceutical, illumination, etc.) | Main types of processed glass: o  lead-alkali-silica ('flint' glass, crystal glass) o  borosilicate glass (good thermal characteristics) o  aluminosilicate glass (good chemical resistance). o  aluminoborsilicate glass (low thermal expansion, high chemical resistance). Produced as flat glass, hollow glass and glass fibre. | **Prompt waste** in glass manufacture:  High level of cullet recovery.  **Process waste** from glass using industries:  High level of recovery of uncontaminated cullet.  **Recovery of post consumer waste** - cullet in domestic and trades wastes (mainly packaging):  generally low level of recovery. |

It is standard practice to utilise waste glass (cullet) in glass manufacture. The raw materials for glass manufacture are mixed with the cullet before entering the melting tank. Cullet melts before the other constituents and thus speeds up the heat circulation through the furnace, and hence the melting, enabling a faster throughput of materials. At present around 20% of glass production in the EEC is from cullet.

**New scrap:** The prompt wastes which arise during the manufacture of glass are extensively utilised. There is virtually no waste from the actual processes employed, but breakages and rejects provide uncontaminated 'in-house' cullet for use in the glass melting operation. Breakages from packaging plants are also a source of cullet. Such 'foreign' cullet (that is cullet from sources external to the immediate production line) is extensively used when it is kept uncontaminated by glass-using industries.

Foreign cullet is used to a greater extent in hollow glass production than flat glass production.

**Old scrap:** The major potential sources of additional cullet for recycling is domestic/trade waste. This originates predominantly from packaging, with the balance consisting of household glassware.

There is also the possibility of reusing glass containers. This was well established practice until the advent of non-returnable bottles. The energy and resource savings available through reuse are considerable.

## 3.6    Rubber

### 3.6.1    Waste arisings and current recovery level

The total arisings of rubber in the EEC in municipal and post-consumer wastes are estimated as 1.6 million tonnes, of which 0.5 million tonnes are currently recovered as a material. The main end uses and current recovery activity for rubber are summarised in Table 3.6(a).

**New scrap:** The consumption of rubber for tyre and tyre-related uses dominates the consumption pattern in Europe. A certain amount of waste rubber is generated in fabrication and, in almost all cases, the scrap can be utilised in-plant either by reprocessing in the same product process line or by down-grading the material for use in a different, less technically stringent, product. Hence, hardly any waste rubber is discharged during normal processing and fabricating operations.

| Table 3.6(a) | | |
| --- | --- | --- |
| RUBBER - MAIN USES AND RECOVERY ACTIVITY | | |
| End Uses | Processed Forms | Current Recovery Activity |
| Tyre covers<br><br>Tyre tubes<br><br>Technical<br>goods<br><br>Hose and<br>tubing | Natural rubber (NR) is<br>produced through tree<br>cultivation.<br><br>Synthetic rubber (SR)<br>production requires<br>the synthesis of<br>monomers and their<br>polymerisation. | **Prompt scrap** arising<br>during NR and SR produc-<br>tion: High level of<br>recovery.<br>**Process scrap** arising<br>during the manufacture<br>of rubber products: High<br>level of recovery. |
| Footwear<br><br>Sports<br>goods<br>Surgical<br>goods<br><br>Belting<br><br>Foam rubber<br>goods<br><br>Adhesives<br>& proofing | Fabricating stages<br>involve:<br>o mixing with<br>chemicals<br>o shaping by forming<br>(extrusion,<br>calendering),<br>moulding, hand or<br>machine building<br>o vulcanisation of<br>the shaped product. | **Post consumer scrap**<br>(rubber tyres): Some<br>recovery (retreading,<br>reclaim and crumb<br>production). |

**Old scrap:** The rubber available for recycling thus appears as post-consumer waste among which discarded tyres are the most prominent element. These end up in a variety of locations including service stations, tyre depots and breakers yards, with a small proportion in domestic wastes.

The other principal source of waste rubber is in domestic and trade wastes where rubber forms a small fraction of the total waste stream. Rubber in domestic wastes is usually soiled and/or considerably mixed with other materials. There is, at present, no mechanical process available to remove the pure rubber content from refuse, particularly from the other non-rubber polymers. The small rubber fraction of domestic waste does not therefore offer any real potential for recycling as a material.

The major recycling opportunity thus rests with used tyres.

## 3.7 Wood

### 3.7.1 Waste arisings and current recovery level

The total arisings of wood waste in the EEC in municipal and post-consumer wastes are estimated at 19.5 million tonnes of which about 0.5 million tonnes are currently recovered. The main end uses and current recovery activity for wood are summarised in Table 3.7(a).

| Table 3.7(a) | | |
|---|---|---|
| WOOD - MAIN USES AND RECOVERY ACTIVITY | | |
| End Uses | Processed Forms | Current Recovery Activity |
| Timber<br><br>Finished articles of secondary industry | Raw timber logs<br>- sawlogs<br>- pulpwood | **Prompt arisings** during logging operations, off-cuts, thinnings, low-grade trees/wood, branches, stumps, roots. Very low recovery levels - some new developments. |
| Architectural woodwork<br><br>Flooring | Pulp | **Prompt arisings** of bark and low grade timber during pulping - high recovery of wood fibres. |
| Furniture<br><br>Plank<br><br>Particle board<br><br>Fibreboard<br><br>Wood by-products, e.g. sawdust, chips and planer shavings | Timber | **Prompt arisings** of bark, off-cuts, peelings, planings, slicings, sawdust, woodflour and woodchip from sawlogs. Recovery levels about 50%.<br><br>**Process arisings** of off-cuts planings, sawdust, woodflour and woodchips. High levels of recovery. |

Wood wastes arise for recovery at a number of points in the sequence of wood processing, namely at:

o       harvest

o       sawmills and

o       secondary converters.

**Timber:** Estimates vary as to the level of wastage within forests but it is likely that about 22% of the total biomass remains after harvesting. Of this 22%, about 25% is available or suitable for recovery as a material or for its energy content.

**Pulpwood:** Conversion of pulpwood logs to pulp results in the production of large quantities of bark which cannot be easily utilised and is therefore available for alternative recovery. About 10 to 15% (weight for weight) of the wood removed is unusable in industry and this is almost entirely composed of bark.

**Saw logs:** Recovery levels for sawmill residues vary widely but a generally accepted level of recovery is around 50%. Sawmill residues are estimated to provide (typically) 35% of the plant's own energy requirements.

Further volumes of wood waste are produced by secondary timber converters and levels of recovery both for energy and materials of between 45 and 60% occur across the Member States.

**Converted wood:** Considerable volumes of waste wood are converted into particle and fibre boards or paper pulp. There is evidence that this market is becoming saturated however and future recovery is more likely to be in the form of energy.

3.8     **Waste Quantities Available for Recovery**

Based on the review of future recovery opportunities for each material, with the exception of small quantities of mixed plastic wastes and process wood wastes, the major opportunities for further material recycling clearly rest with municipal and other post-consumer wastes. The levels of arisings, quantities currently recovered and wastes still physically available for recovery are given in Tables 3.8(a) and 3.8(b).

In addition to the municipal and post-consumer wastes available for further recovery it is estimated 200,000 tonnes of mixed plastics and 6 million tonnes of wood are also available as process wastes.

| Table 3.8(a) | | | |
|---|---|---|---|
| ESTIMATED WASTE ARISINGS AND CURRENT RECOVERY LEVELS MUNICIPAL WASTES (All Data Mt) | | | |
| | Estimated Arisings (1) | Currently Recovered (2) | Physically Available for Further Recovery |
| Aluminium | 0.9 | *(3) | 0.9 |
| Plastics | 4.5 | 1.0 | 3.5 |
| Waste Paper | 28.1 | 12.0 | 16.1 |
| Glass | 8.9 | 1.8 | 7.1 |
| Rubber | 1.0 | 0.2 | 0.8 |
| Wood | 1.5 | 0.5 | 1.0 |
| (1) Data sources given in Appendix A. (2) Excludes secondary materials recovered as energy or compost (3) * denotes negligible | | | |

Owing to the lack of reliable data a degree of estimation is
necessary in deriving quantitative data on arisings and recovery
levels for different categories of wastes.  A wide variety of
sources have been used in arriving at the figures shown in the
tables but it must be recognised that these are estimates and
subject to a margin of error.

| Table 3.8(b) | | | |
|---|---|---|---|
| ESTIMATED WASTE ARISINGS AND CURRENT RECOVERY LEVELS POST-CONSUMER WASTES (All data Mt) | | | |
| | Estimated Arisings | Currently Recovered | Physically Available for Further Recovery |
| Aluminium | 0.6 | 0.4 | 0.2 |
| Plastics | 7.1 | 0.7 | 6.4 |
| Waste Paper | *(1) | * | * |
| Glass | * | * | * |
| Rubber | 0.7 | 0.3 | 0.4 |
| Wood | 18.0 | - | 18.0 |
| (1) * denotes negligible | | | |

The quantities of wastes denoted as physically available for further recovery (right hand column on the tables) represent the theoretical maximum. Owing to factors such as geographical dispersion and contamination not all these wastes can be realistically considered as substitutes for primary materials. Further adjustments are therefore necessary to derive quantities of secondary materials which can be considered as practically available for recovery for use in energy saving calculations. This is undertaken in Section 7 after considering alternative recovery approaches.

**4. METHODS OF RECOVERING MATERIALS AND ENERGY FROM REFUSE**

**4.1 Introduction**

The waste materials arising in municipal refuse are available for recovery either as secondary materials or as feedstock energy. In the case of recovery as a material there is the further choice of separation at source or mechanical separation. These different options are described below. This is followed by an assessment of the factors affecting the viability of further recovery.

**4.2 Material Recovery**

In the previous section the main area of potential for the additional recovery of materials was identified as being domestic and trade wastes. The crucial question is how, and to what extent, recovery from these sources can be achieved. There are two main options, although each has several variants:

o    mechanical separation of materials from mixed refuse at centralised locations;

o    material sorting at source by the consumer.

A review of mechanical and source separation schemes is given in Appendix C. In this section we summarise the main features of these alternative approaches and assess to what extent additional material recovery is likely to be achieved by these different routes.

**4.2.1 Source separation schemes [1]**

This is a well established method of recycling materials and can take several forms, but there are two basic approaches which can be adopted:

o    waste separated at source for collection by a third party;

o    waste separated at source and deposited by the consumer in a neighbourhood or centrally located container for collection by a third party.

---

[1] Source separation has been defined by the US Environmental Protection Agency as: "the setting aside of recyclable waste materials at their point of generation for segregated collection and transport to specialised waste processing sites or final manufacturing markets. Transportation can be provided either by the waste generator, by city collection vehicles, by private haulers and scrap dealers, or by voluntary recycling or service organisations."

The recovery of paper and glass by these methods is well established in the EEC and more recently the recovery of cans and plastics by similar methods is being undertaken. Certain general observations can be made on the operational experience of these schemes, as summarised below and in Table 4.2(a):

o   **the product to be returned must be easily identifiable** otherwise large quantities of unsuitable material will be included, requiring separation and subsequent disposal. Thus, for instance, it is necessary for the consumer to be able to distinguish between aluminium and tinplate cans and between PET and PVC bottles;

o   **the recovered materials are relatively uncontaminated.** Apart from closures, labels, varnishes etc., used in packaging materials and containers, the only other contaminant is usually residual liquid. However, the potential for recycling materials may be considerably enhanced or severely diminished according to how well householders handle, process and store potentially recyclable materials;

o   **it is not yet clear whether consumer motivation can be sustained over long periods of time** (with or without payment). Recovery of glass and paper has traditionally relied on good-will and an appreciation of environmental and conservation arguments to gain cooperation. The recovery of aluminium cans however is based on the American approach of paying for cans returned. This approach is only possible with a high value material and it is doubtful whether it is appropriate for the other materials being considered. Thus the participation rates which can be achieved and sustained for difficult product groups have yet to be established. An extensive and continuous education campaign will almost certainly be a requisite of maintaining and increasing consumer participation;

o   **collection costs (dominated as they are by labour costs) are likely to be substantially less if voluntary groups perform the collection activity** as opposed to local authority collection crews. However, the transporting of small quantities of materials by many individuals is likely to be less efficient in conserving both energy and other environmental resources than collection by a single agent. The key point here concerns the location of the central collection facility. The central containers or collection centre must be located in positions that would be passed or visited by the collectors in the course of some other normal activity, such as shopping expeditions. The recyclable materials, can then be economically transported to the site which jointly serves as a recycling location and, say, a shopping location.

Table 4.2(a)

COLLECTION OF SOURCE SEPARATION MATERIALS - ADVANTAGES AND DISADVANTAGES

| Type of Collection System | Common Advantages | Common Disadvantages |
|---|---|---|
| **Central collection point systems** Drop-off centres and/or transfer containers; 'conveniently' positioned containers or skips to which waste generators take their paper, glass or plastics residuals; containers can be in 'neighbourhood' sites for multi-family dwellings or more centralised locations such as supermarket carparks. | Minimum investment in plant and manpower. Flexibility of collection frequency and relatively low collection costs. Drop-off centres often expand into more extensive schemes involving collection rounds or networks of centres. | Possible inconvenience for participants and thus risk of lack of long-term participation co-operation. Lack of control over quality/ contamination levels. Possible vandalism and site environmental problems. Capital costs of the more sophisticated types may prove burden-some. |
| Recycling centres; buy-back schemes. Participants bring residuals to a centralised handling/processing facility. Sometimes combined with door-to-door collection by agency running the centre. | Good quality/contamination control possible; social benefits (community involvement); participation encouraged by payments for higher value residuals. | Participant inconvenience and possible co-operation problems especially if no payments are made. |
| **Collection round systems** Combined door-to-door, collection of general refuse and separated materials utilising various vehicle types (e.g. rack or box systems, trailer system, compartmentalised truck, articulated container carriage). | Simultaneous collection should aid partici-pation rates; rack or box system has low capital cost and entails only a minimum disturbance of the existing system. The trailer option yields capacity advantages and minimum system disturbance. The compartmentalised vehicle involves minimum additional collection time at each pick-up - collection costs per tonne lower. Articulated containers facilitate unloading at transfer station. | Relatively high overall collection costs; rack system has capacity problems; trailer system increases loading time (interference with rear-loading operation) and has movement and manoeuvring problems (worker safety). Purpose-built vehicles involve high capital costs. Transport cost problems if disposal facility and recycling station not conveniently located. |
| Separate door-to-door collection with a variety of possible vehicle container types. | Flexibility of collection frequency. No manoeuvring problems. Quality control possible (except when opaque plastic bags are utilised). | Capital cost of vehicles (unless volume of recyclables diverted from mixed refuse collection is large enough to justify reducing trucks and crews on this service). Collection of recycl-ables on a day other than regular refuse collec-tion often makes the collection schedule confusing to residents and as a result partici-pation rates may fall. Relatively high collec-tion costs. |

Source: ERL and Source Separation and Recycling Schemes; R.K. Turner and C. Thomas, Resources Policy, March 1982.

**4.2.2     Mechanical separation schemes**

A wide variety of recycling methods have been developed in recent
years to reclaim components of mixed wastes by means of automatic
sorting processes.  Many of these have not been developed beyond
the pilot and experimental plant stages and there is very limited
experience of large scale 'commercial' plant operations.  In
Appendix C we review the few plants which are engaged in material
recovery activities at present.  The main points to be noted are
as follows:

o         The technical potential exists for the mechanical
          separation of paper, glass and aluminium.  The separated
          materials however are generally of low quality and
          manual separation is often additionally required.  The
          mechanical separation of plastic (apart from plastic
          film) is still in the development stage.  Information on
          the actual technical performance of operating plants is
          not publicly available.

o         There is no reliable cost and revenue data available for
          the plants currently operating.  All were built with
          financial assistance from government sources and it is
          not clear to what extent on-going operational subsidies
          (open or hidden) are provided.

o         Mechanical separation plants are only likely to be
          viable in urban areas where a sufficient volume of
          wastes can be generated to provide an economic base
          load.

o         The inhomogenity of the feed and fluctuating marketing
          conditions require flexibility to be built into the
          plant's design.  Thus the ability to switch between
          recovery of different materials and between materials
          and RDF in response to market conditions provides
          greater opportunity for maximising economic performance.

o         As the output of separation plants needs to be marketed,
          their management should be in the hands of operators
          with a good knowledge of secondary markets and with the
          entrepreneurial talents to respond quickly to market
          opportunities.  This may require distancing these
          operations from the direct control of the waste disposal
          authority.

**4.2.3     Source separation and mechanical recovery compared**

The main point of interest in respect of energy analysis is the
extent to which further recovery will be achieved through source
separation as compared to mechanical separation.  In Table 4.2(b)
we review both recovery systems against key parameters.

The two systems are not mutually exclusive. Mechanical separation can separate out the materials remaining after source separation but this inevitably leads to both a lower quality feedstock and reduced volume throughput.

Table 4.2(b)

| Item | Source Separation | Mechanical Separation |
|---|---|---|
| Technical Feasibility | Relatively simple. | Sophisticated equipment required. |
| Quality of Material | Reasonable quality and limited additional separation. Good opportunity for recycling material back to original use. | Low quality generally and considerable reprocessing required. Limited opportunities for recycling material back to original use. |
| Market Acceptability | Reasonable. | Poor. |
| Reliability of Supply | Reasonable up to a certain level of recovery. Continuing educational/information programme required to maintain consumers' cooperation. | Variable at current level of technology. Also subject to fluctuation when material prices rise and source separation activity increases. |
| Investment Costs | Variable, depending on type of schemes. Low if high degree of collection by voluntary groups or by consumers' own transportation. | High capital investment. |
| Economic Viability | Variable, depending on type of material and price levels in secondary market. | Not yet proven for large scale operation. Entrepreneurial approach required. |
| Organisational | Involvement of final users of recovered materials or professional recyclers preferable. | Involvement of secondary market specialists in management of plant. |

The extent to which mechanical separation can become established as an economically acceptable and technically reliable form of materials recovery is still very uncertain. Given the degree of development work still being undertaken in Europe, supported by national governments and the European Commission, it is probable that technical advances will be made; whether this can be achieved at an acceptable cost remains to be seen. The great advantage that mechanical separation enjoys over source separation is that all materials pass through the system; source separation will always depend on the cooperation of the public - a rather uncertain commodity.

On the basis of current level of technology, however, source separation would seem the most viable option for recovery of the materials being considered. Aluminium, plastics, paper and glass are all suitable for such an approach. Wood and rubber do not arise in significant quantities from these sources and can be ignored.

## 4.3    Energy Recovery

The alternative to the recycling of combustible materials (plastics, paper and rubber) is to burn them for their direct fuel value.

Energy has been recovered for many years from untreated domestic refuse by way of the hot combustion gases generated by direct incineration methods. More recent approaches to combustion of municipal waste with energy recovery have been to convert raw refuse into more conveniently handled fuels in solid form. Refuse Derived Fuel (RDF), a product which stores the energy value of municipal waste for burning, is the area in which most development work has been undertaken. But there are other developments in the pyrolysis and gasification of wastes which are also receiving considerable attention; these processes provide gases, oil and char which have potential uses as fuel.

Below we summarise the current status in the exploitation of these developments and of traditional incineration.

### 4.3.1    Incineration

Conventional Incineration

In Europe and the USA refuse incineration is a well-established disposal technique. A review of refuse incineration was contained in 'Energy from Municipal Waste - 1980/81, Summary Report, CEC' and it is not proposed to include a detailed exposition here.

Obviously, the extent to which paper and other waste streams are combusted makes them unavailable for reuse as secondary materials, although such recovery is not always technically possible.

In 1981, about 24% of all municipal waste was incinerated, but only 65% or about 15.0 million tonnes of this amount was incinerated with heat recovery, i.e. energy was recovered. This is unevenly distributed among Member States with the majority residing in Germany and the Netherlands. More recently, energy recovery capacity has been growing in Italy.

There are two main options for direct (on-site) energy recovery associated with incineration:

(i)    for **heat** by producing steam, and distribution to consumers as low grade heat for district heating (or for industrial use);

(ii)   for **electricity**, with a conventional boiler turbine for local distribution or sale into the national grid.

Energy recovery from municipal waste is a capital-intensive process. The operation and maintenance of the plant requires skilled manpower and plants have often not achieved their design throughput capacity because of unscheduled maintenance shutdowns. The overall economics of incineration with energy recovery, and therefore its feasibility, depends critically upon the price received for the energy (usually electricity) produced.

### Starved Air Combustion

Starved air combustion of wastes differs from conventional incineration by restricting to a minimum the amount of air necessary to support initial combustion. If necessary, auxiliary fuel is supplied (gas or oil) to complete the combustion process. The main advantage lies in the lower volume of exhaust gases produced which can be more easily controlled.

Whilst a few plants have been installed in Europe for incineration of municipal solid waste in small municipalities, none have included heat recovery, auxiliary fuel generally being required to support combustion. However, a number of plants have been installed for incineration of high calorific value industrial wastes (paper, plastic packaging, timber, etc.) including heat recovery.

The vast majority of the operational plants are in the United States.

### 4.3.2    Refuse derived fuel

### Production

There are currently several full-scale RDF manufacturing plants (mainly pelletised municipal waste) in Europe and North America. These have generally been funded by central government grants, so the development of the RDF process might not yet be considered fully "commercialised". However, considerable experience has been gained in the production of RDF in various forms, where various non-combustible and putrescible fractions of waste are separated.

Refuse derived fuel (RDF) is designed to:

o        improve the calorific value and combustion characteristics of the waste;

o        improve its handling characteristics so that it can be
burnt in conventional solid fuel boilers;

o        reduce ash disposal and other residue problems.

In market terms, RDF is basically competing with coal. However,
different applications demand different quality products.
Technically it is possible to produce highly refined fuel
products from refuse. However, this clearly results in higher
production costs, including energy inputs, higher levels of
reject material for disposal and a higher sale value.

Types of fuel

There are five basic types of RDF that have been developed to
date. These are:

o        Coarse
o        Fluff
o        Densified
o        Powder, and
o        Pulped.

Developments in Europe have concentrated on coarse and densified
fuel products. Developments of fluff, powder and pulped fuels
have taken place mainly in the United States but have now been
largely abandoned. The main characteristics of these fuels are
summarised in Table 4.3(a):

4.3.3     **Pyrolysis/gasification**

Processes

**Pyrolysis** of organic raw materials is a thermal degradation
process carried out in the absence of oxygen. The reaction
conditions (e.g. temperature, pressure, residence time of the
solid and volatile material) and the composition of the feedstock
will determine the relative amounts of gaseous and liquid
products and carbonaceous residue produced.

**Gasification** of refuse is a process whereby the thermally
decomposing material and its carbonaceous residues are allowed to
react with gases such as air, oxygen, steam, carbon dioxide or
hydogen. In the case of air, oxygen or hydrogen the reaction
with the gasified material is exothermic and the heat generated
can be used to obtain or maintain the desired reaction
temperature.

| Table 4.3(a) |
| --- |
| MAIN TYPES OF RDF |

**Coarse**

**Single shred** - refuse is processed by size reduction (to less
than 100 mm) using hammer or flail milling, mill or drum pulverisation or shredding, with the magnetic separation of ferrous
metals to produce a more homogeneous material for firing. The
calorific value of the fuel produced by this process from
municipal solid waste is typically in the range 9.30-10.50 GJ/t.

**Double shred** - in order to achieve a smaller particle sized RDF
(typically less than 60 mm) to assist in handling and firing, a
second shredding stage is included. The typical calorific value
is in the same range as the single shredded product (9.30-10.50
GJ/t).

**Fluff**

This product results from a process which includes air classifi-
cation, giving a 'light fraction' (comprising mainly paper and
plastic from municipal solid waste) with a higher calorific value
fuel (typically 11.60-12.80 GJ/t).

Both coarse and fluff RDF are costly to transport due to their
low bulk density and are difficult to store as they have a
tendency to ferment and compact. They can also present a fire
hazard due to spontaneous combustion. These problems have led to
the development of the following product processes:

**Densified**

This category includes a group of different processes whereby the
'fluff' material is extruded or pressed into cubes, pellets or
briquettes for improving handling and transport and storage
properties. Typical calorific values are in the range 11.60-
15.00 GJ/t.

To achieve consistent standards of pellet quality a drying stage
is often included. This results in a higher calorific value
product, but requires a higher process energy input.

**Pulped**

Under this process refuse is fed to a hydrapulper to produce an
aqueous slurry from which the heavy fraction is removed and the
light fraction is reduced to a 50% moisture content fuel. The
latter has a calorific value of 8.00 GJ/t.

Technical Status

The few plants now in operation in Europe, of which only one is a full-scale pyrolysis unit at Creteil in France, represent the third generation of pyrolysis/gasification process technology. There are one or two other laboratory scale systems operating and two demonstration plants. Previous generation pyrolysis plants failed on technical/economic grounds.

In the Federal Republic of Germany, experience with two plants (one pilot and one larger demonstration plant) has led to a renewed interest in pyrolysis where the technology is seen as offering a particularly appropriate and potentially economic solid waste treatment/energy recovery process for certain medium sized municipalities. The performance of the two plants, BKMI and Kiener rotary kiln processes at Aalen and Günzburg, will be critical in influencing the future of pyrolysis as a viable energy recovery technology for municipal waste.

The principal problems with previous pyrolysis plants were associated with the unreliability of the process caused by slagging in the reactor and bridging and chanelling of the refuse feed. The new generation of rotary kiln units may overcome some of these difficulties.

The process is also being applied to high calorific value industrial waste such as tyres, plastics and contaminated paint solvents/ wastes.

Fuel Produced and Applications

The fuel products available from pyrolysis/gasification processes and the associated energy values are as follows:

**Gases:** The calorific value and available quantities of pyrolysis gas are such as to make it uneconomic to transport over long distances either by compression or liquefaction. It is therefore most suitable for use for on-site steam raising either for plant/district heating or electricity generation. Whilst, in theory, other possible uses exist as a synthesis gas in chemical processing or, following methanation as a natural gas substitute, the quantities involved are unlikely to be sufficient to justify its use in such applications.

**Oils & Tars:** Tars and heavy oils are likely to find outlets as fuel in the pyrolysis process, or as low-grade refinables to be collected with other waste oils.

**Char/Solid Residue:** There are three theoretical potential uses for the char:

o        as a substitute for commercial activated carbon;

o        as a fuel;

o        as a substitute for sand or gravel in road construction.

The first of these is unlikely to be worthwhile due to the further processing required to upgrade the product for use as an adsorbent.

The high ash content and relatively low calorific value make its use as a fuel of only marginal interest.

The granulated slag from the high temperature processes may find a market in road construction.

## 4.4.    Factors Affecting Further Recovery

### 4.4.1    Markets for Recovered Materials

The extent to which the amount of waste material still physically recoverable can actually be recycled depends on a number of factors. A vital consideration is whether the material can be recovered in a form and at a quality and price for which demand exists. In Table 4.4(a) the sources, possible degrees of contamination and end uses of the materials under review are summarised. The main uses are as follows:

### Aluminium

Recovered aluminium is remelted for use as a direct substitute for primary aluminium across a wide range of uses. Recovered all-aluminium cans can be reprocessed as sheet for reuse in can production. Aluminium recovered from capital goods is similarly capable of being reused across a wide spectrum of uses, the precise nature depending upon its grading (purity).

### Plastics

The reuses of mixed plastics arising in municipal wastes are limited by the current level of technology (see Section 3.3 and Appendix B), which cannot separate these wastes into separate polymer types. When mixed wastes are recovered they are recycled into low grade, non-critical, products where they often act as substitutes for other materials.

The further recycling of mixed plastics (apart from as a fuel or chemical) is thus dependent on the development and exploitation of new products.

Such products include solid tyres, flooring and articles for agricultural/horticultural applications requiring toughness and corrosion resistance such as planks, sheets, posts, grids, fencing and animal housing components.

Where there is a high volume of a specific polymer in the wastes, recovery may be possible as a plastic substitute. This occurs at the Rome separation plant for instance where polyethylene is separated out for reprocessing into waste sacks. Also, PVC bottles used for mineral water are being collected in France for use in the manufacture of pipes and conduits for cables.

Table 4.4(a)

SECONDARY MATERIAL SOURCES AND END USES

| Type of Secondary Material | Grade (value) | | Possible Contaminants | | | End Uses | General Source |
|---|---|---|---|---|---|---|---|
| | Degree of Homogeneity | Degree of Contamination | During Manufacture | During Usage | During Handling After Use | | |
| **Aluminium** | | | | | | | |
| Offcuts, turnings & rejections | High | Low | High quality - no contaminants | | | Reused in new aluminium products | Metal fabricating plants, can makers |
| Used aluminium cans, foil and bottle tops | Fairly high | Fairly high | | Paper labels contents of can | Dirt, fats, papers, plastics | Resmelted into ingots to make new cans | Households |
| Post consumer aluminium in capital goods particularly cars | Variable | Variable | Alloyed | Plastics and non-ferrous metals | | Resmelted into ingots | Post Consumer waste streams |
| **Plastics** | | | | | | | |
| Wide diversity of plastic materials ranging from thin polyethylene film to solid plastics and polypropylene twine: wide range of product forms, offcuts, crates, buckets and other containers, sacks etc. | Variable heterogeneity often is problem | Variable | | Paper labels, metal labels, lids and foil content of bags or sacks, ink and gum | Adhesive tapes, wood, dirt chemicals | Reused in pelletised form: sacks, bags, pellet shroud, building products, crates and other containers, filler products | Polymer manufacturers, converter plants, packaging manufacturers, carpet manufacturers, plastics-coating processors, farms etc. |
| Thirty different types of plastic in everyday use: solid plastics, containers, polyamides, PVC, shrink wrap plastic polyethylene tere-phthalate (PET) bottles etc. | Variable often low | Potentially high | | | | Mixed plastics can form part of refuse-derived fuel pellets; PET bottles can be used as filling fibre products - fillers for sleeping bags or quilted anoraks | Households, small retail outlets |

continued/...

Table 4.4(a) continued
SECONDARY MATERIAL SOURCES AND END USES

| Type of Secondary Material | Grade (value) Degree of Homogeneity | Degree of Contamination | Possible Contaminants During Manufacture | During Usage | During Handling After Use | End Uses | General Source |
|---|---|---|---|---|---|---|---|
| **Waste Paper & Board** Clean unprinted papers; printers waste, paper converter paste, stationer and book binder waste etc. | High | Low Grades 1 & 2 (paper making grades) Grades 1-4 (paper making grades) Grades 5-7 (board making grades) | Printing inks, plastic coating, bitumen wax, laminates, wet strength resins, latex (rubber) | Poly-styrene foam, staples, pressure-sensitive glues, sealing tapes, wire, carbon paper, product contents of containers | Wire staples, paper clips, metal and plastics, string, twine, stones, grit, foil, dirt, rubber bands | Paper making, high quality products; liners for boards, printing and writing papers | Industrial/converting sources: printers, converters, news-paper publishers |
| Tabulating cards and listings, envelope papers, office shavings, office papers (forms, ledgers, etc). | High Grades 3 & 4 | Low | | | | Tissues, wrapping paper, liners and envelope papers; deinked office papers can be used for printing and writing papers and tissues | Large industrial/commercial sources; institutional sources |
| Newspapers, brochures and directories; over issues, returns and used newspapers from households | Fairly high Grade 5 | Possible contamination in old news-papers | | | | Liners for board, after deinking old newspaper can be used for newsprint, cellulose insulation products, animal bedding, smoke-less fuel blocks, moulded products | Newspaper and magazine printers; households |
| Corrugated paper-board waste; kraft paper | Fairly high Grade 6 | Possible contamination in shop waste | | | | Brown paper bags, sacks, packaging boards, boxes, cartons & cardboard | Industrial sources, sack bag and box manufacturers; shops |
| Mixed papers and board | Low Grade 7 | Potentially high | | | | Paperboard production; building board refuse derived fuel products when combined with plastic waste | Shops, schools, small offices and households; board converters |

continued/...

Table 4.4(a) continued
SECONDARY MATERIAL SOURCES AND END USES

| Type of Secondary Material | Grade (value) | | Possible Contaminants | | | End Uses | General Source |
|---|---|---|---|---|---|---|---|
| | Degree of Homogeneity | Degree of Contamination | During Manufacture | During Usage | During Handling After Use | | |
| **Waste Glass** | | | | | | | |
| Clean glass rejects, breakages, process cullet | High | Low — High Grade | Metal caps and lids, plate and technical glass | Paper, labels, contents of containers | Ceramic materials porcelain stones paper | Reused in glass melting process; glass fibre, ballotini, abrasives, tiles | Industrial glassworks, container plants |
| Filling/bottling cullet (bottles and jars) | Fairly high | Fairly low — High Grade | | | | Similar to above | Bottling, processing and packaging plants |
| Semi-industrial cullet (mainly bottles) | Fairly high if sorted | Fairly low — Middle Grade | | | | Similar to above | Pubs, hotels, restaurants |
| Mixed cullet (green, clear and amber coloured glass) mainly glass containers | Low colour sorting required | Potentially high — Low Grade | | | | Green and amber glass containers, road surfacing material, construction material, wall and floor materials, bulk filler for paint | Households, small shops |
| **Rubber** | | | | | | | |
| Used tyres | High | Low | | | | Retreading 'reclaim' for tyre carcass compounds, fan belts hose, crumb for carpet underlay and road surfacing | Tyre dealers, service stations breakers yards |

In the UK an experimental scheme for the collection of PET
bottles is in operation to determine the possibility of recycling
either as:

o        a fibre for use in sleeping bags;

o        industrial strapping tape; or

o        extruding as a film for insulation curtains.

These applications are still being explored and are not yet
established as commercial operations.  Other uses being developed
in the USA include kitchen scouring pads, floor tiles,
unsaturated polyester and plastic foam.

Unlike aluminium and glass, plastic bottles cannot be recycled
back to their original use because of residual traces.  These
adversely affect the degree of purity required for food and
beverage packaging.

There are opportunities for recovery from other post-consumer
sources when a single polymer is involved.  Thus there are good
recovery prospects for the polypropylene used in battery cases.
Also shredding techniques are developing to isolate plastics in
other car components.  Recovery from electrical cables and
telephone sets is also well-established.

A further source of distribution waste available for the recovery
is discarded packaging, which can be kept apart and collected
separately.  This includes polyethylene fertiliser sacks, shrink-
wrap film and other polyethylene film discarded by supermarkets
and warehouses, bottle crates and food trays.

Another example of direct recycling is taking place in the UK
with the recovery of plastic coat hangers mainly from stores and
clothing manufacturers.  The plastics (polystyrene and polypropylene) are recompounded and reused in the production of new
hangers.

In the EEC more than one billion crates and trays have been in
regular use since their introduction fifteen years ago in
bakeries, dairies, breweries, agriculture, horticulture, supermarkets, etc.  When damaged or at the end of their life they may
be reground and combined with virgin polymer for the production
of new crates.

**Paper**

There are sharp differences in the extent to which waste paper
is, and can be, used in different paper and board products.  The
maximum proportion of waste paper which can be recycled in
different types of paper and board are shown in Table B3.3(a) in
Appendix B.

There is considerable development work being undertaken to overcome the technical constraints to further recycling. This centres around improving waste paper quality by the removal of impurities (or 'contraries'). Even assuming progress is made in this area, and assuming that existing industry and consumer resistance to using products with a recycled paper content can be overcome, there is a need for other uses to be identified as the paper and board industry cannot absorb all the arisings from municipal wastes. Certain areas are being explored but at present there is no immediate prospect of a large market development in new products for low grade waste paper.

## Glass

The use of outside or 'foreign' cullet by glass manufacturers is the main outlet for reclaimed glass. Only small quantities are used for by-products - such as road surfacing materials, wall and floor materials, construction materials and bulk fillers for paints. Further use of cullet will depend on higher utilisation in existing uses and the development of new markets. It is possible that additional cullet could be utilised by glass manufacturers for certain products without sacrifice of quality, subject to a relative cost advantage of cullet compared to glass raw materials.

## Rubber

Apart from the recycling of old tyres for retreading the main markets for recovered rubber are as rubber 'reclaim' and rubber crumb (powder).

The potential exists for increasing the number of retreads for car tyres; this will require, among other things, changes in consumer attitudes to retreaded tyres and a sufficiently attractive price differential between new and retreaded car tyres.

The use of reclaim (chemically treated vulcanised rubber) both in tyres and non-tyre applications has been declining in recent years. It is recognised that new markets need to be found or created if the full production potential of reclaim is to be realised. Research and development programmes are being carried out in the USA and Europe into new applications.

Rubber crumb (ground vulcanised rubber) is used in the manufacture of carpet underlay, road surface applications and shoe soles. As with reclaim, new markets are required if the recycling potential is to be realised.

## Wood

The markets for wood waste materials are highly dependent on the nature and homogeneity of the material. High grade softwood fibre residues have a number of potential uses as pulp substitute and in production of wood particle boards. Mixed wastes are constrained by their poor quality and unless separation is economic, fibreboard remains the main market for these materials.

Bark has limited use apart from the conversion chemically to phenolic resins and adhesives and direct use horticulturally or agriculturally.

A great deal of research is currently underway to find new uses for wood waste as a material. Further recovery could be greatly constrained as processing units and markets are geographically distant.

### 4.4.2  Markets for Fuels

Potential users of the fuel or energy produced from refuse fall broadly into four groups:

o       industrial or commercial energy consumers with boilers and kilns based on solid fuel (usually coal) whose fuel systems might be suitable for various alternative cheaper fuels based on solid wastes;

o       industrial or commercial users who can adapt boilers or power plants to run on gas produced by pyrolysis/ gasification plants;

o       industrial, commercial or domestic users of hot water/ steam for process/space heating who might be interested in being supplied via a district heating network;

o       utility companies who would purchase purified gas or surplus electricity for trunk/grid distribution.

RDF has a higher ash and moisture content than coal and a lower calorific value. Existing boilers designed for coal firing may therefore need modification and additional operating costs, including labour, may be incurred. The price of RDF must be sufficiently lower to more than offset any such cost disadvantages. Reliability of supply will also be an important factor, together with adequate controls on product quality.

Solid fuel boiler operators, raising steam and hot water for a variety of process and space heating purposes have the following overall objectives:

o       to minimise total energy utilisation costs, including fuel preparation, handling, ash disposal and boiler maintenance costs, as well as fuel firing;

o       incorporated within the above, not to add significantly to manpower costs, either in terms of numbers of operators or level of skills required;

o       trouble-free operation;

o       reliability of fuel supply;

o       to meet any legal requirements of local air emission control.

Users of energy recovered from municipal wastes will need to be located close to the waste/energy recovery plant facility to make supply feasible and economic.

Thus it may be seen that while fuel derived from municipal solid waste clearly offers a potentially valuable fuel supply to industry, it must be marketed by the producers in such a way and at such a price that it provides an attractive economic and operational alternative energy source for the user.

In some instances, producer and user will be the same organisation and this has associated benefits in both cost and organisational terms. It may also prove advantageous in removing the energy consumption of fuel delivery to consumers from the energy balance equation.

### 4.4.3 Other Factors Affecting Recovery

There are a number of other factors which influence the viability of recovery schemes, as summarised in Table 4.4(b). But in the last resort increased demand for secondary materials will depend on factors such as:

o the technical barriers to substitution and relative prices of secondary and primary materials;

o industry and consumer attitudes to products containing recovered materials;

o reliability of supply and consistency of quality;

o the development of new markets. As noted in 4.4.1 above, all the materials, with the exception of aluminium, are dependent on the development of new applications for realising their maximum recovery potential.

For a given level of technology it can be argued that, in the absence of government or Community intervention, the market will be close to equilibrium and recycling at its optimum level (in financial cost terms). If this is the case, further recycling will not automatically occur even if additional material is recovered. The mere existence of a supply of recovered material will not create a demand for it.

Thus while a potentially 'available' quantity of material can be estimated for further recovery there are many technical, organisational, market and financial reasons why this may be unattainable. To the extent that governments consider it desirable to promote further recovery for reasons of material and energy conservation (or other reasons) it can intervene, directly or indirectly, to help bring this about.

For the purposes of estimating energy savings therefore, no consideration has been given to the costs of further recovery or whether markets exist for the recovered materials at present.

The energy savings thus calculated represent the maximum attainable.

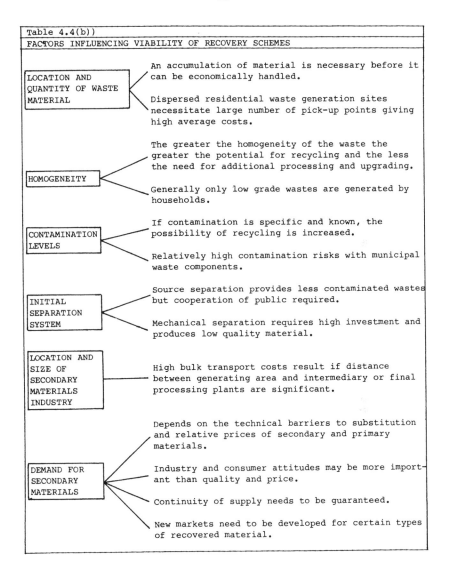

Table 4.4(b))

FACTORS INFLUENCING VIABILITY OF RECOVERY SCHEMES

LOCATION AND QUANTITY OF WASTE MATERIAL

An accumulation of material is necessary before it can be economically handled.

Dispersed residential waste generation sites necessitate large number of pick-up points giving high average costs.

HOMOGENEITY

The greater the homogeneity of the waste the greater the potential for recycling and the less the need for additional processing and upgrading.

Generally only low grade wastes are generated by households.

CONTAMINATION LEVELS

If contamination is specific and known, the possibility of recycling is increased.

Relatively high contamination risks with municipal waste components.

INITIAL SEPARATION SYSTEM

Source separation provides less contaminated wastes but cooperation of public required.

Mechanical separation requires high investment and produces low quality material.

LOCATION AND SIZE OF SECONDARY MATERIALS INDUSTRY

High bulk transport costs result if distance between generating area and intermediary or final processing plants are significant.

DEMAND FOR SECONDARY MATERIALS

Depends on the technical barriers to substitution and relative prices of secondary and primary materials.

Industry and consumer attitudes may be more important than quality and price.

Continuity of supply needs to be guaranteed.

New markets need to be developed for certain types of recovered material.

5.     **UNIT ENERGY SAVINGS RESULTING FROM RECOVERY OF MATERIALS FOR RECYCLING**

5.1    Introduction

Energy savings result from waste recycling if the energy used in collecting, separating and treating reclaimed wastes, and subsequent processing, is less than the energy used in originating and processing primary materials and disposing of wastes.  The different stages involved are shown in Table 5.1(a).

| Table 5.1(a) |
| --- |
| DIFFERENT STAGES AT WHICH ENERGY CONSUMED IN PRIMARY AND SECONDARY MATERIAL PRODUCTION AND USE |

**PRIMARY MATERIAL PRODUCTION/USE**

Stage 1:     Extraction of raw materials

Stage 2:     Transportation of raw materials to primary production unit

Stage 3:     Production of primary material

Stage 4:     Transportation of primary material to product manufacturing unit

Stage 5:     Product manufacture

Stage 6:     Product distribution and consumer use

Stage 7:     Waste collection/disposal/recovery

**SECONDARY MATERIAL PRODUCTION/USE**

Stage 1:     Collection of waste material

Stage 2:     Sorting and pre-preparation of material

Stage 3:     Transportation to secondary material production unit or use as fuel

Stage 4:     Production of secondary material

Stage 5:     Transportation of secondary material to product manufacturing unit

Stage 6:     Product manufacture

Stage 7:     Product distribution and consumer use

Stage 8:     Waste collection/disposal/recovery

58

The comparison of energy consumption associated with the primary
and secondary material production/use cycles is made by comparing
the aggregates of Stages 1 to 3 in primary production with Stages
1 to 4 in secondary prodution.  In addition, any savings in
disposal energy in the primary cycle are credited to secondary
materials.  Stages 4 to 6 in primary production and Stages 5 to 7
in secondary production are common to both cycles and can
generally be ignored for comparative purposes.  The retreading of
rubber tyres  is the one exception where different energies in
product manufacture arise.

There are considerable variations in material production
processes both within and between countries; these depend on the
purity and form of primary or secondary materials used.
Consequently, the potential energy savings resulting from primary
material substitution are wide ranging.  The efficiency and age
of plant also affect processing energy requirements.

Below we review the energy savings which can typically be
achieved through primary material substitution.  For the reasons
just noted, no two situations are identical but a representative
percentage energy saving figure has been derived for use in
calculating overall energy savings.

We also review the typical energies involved in the transportation of materials and waste disposal.


5.2     **Aluminium**

5.2.1   Introduction

The smelting of alumina (refined bauxite) into aluminium ingots
is a highly energy-intensive process, using mainly electric power
for electrolysis.  Whereas the raw material input represents
typically about a quarter of the final price of an ingot of
aluminium, energy costs account for up to 55% of total costs.

There have been steady improvements in smelting technology in
recent years with energy savings of 1-2% per annum being achieved
over the last ten years [1].

Two points should be noted regarding the potential for energy
savings to be achieved through substitution of aluminium for
primary raw materials:

    i.      a considerable proportion of aluminium ingots processed
            in the Community are imported from low electricty cost/
            energy rich countries;

[1]  Developments in the Energy Utilisation of the Primary
Aluminium Industry, M.D. Lester.  UNEP Industry & Environment, 6,
3, 1984. (in press)

ii.    the amount of primary energy saved depends on whether the consumer electric power is generated thermally or from hydro-power.

It is therefore difficult to be precise about the actual amount of energy saved through substitution of secondary aluminium.

Energy efficiencies vary from plant to plant and area to area. Such variations reflect the different ages of smelters, which are costly to modify and modernise once installed, and the relative energy endowments of different countries and regions and methods of electricity generation: smelters in energy-rich areas tend to be less energy efficient.

Typical energies involved in producing aluminium from primary and secondary materials are reviewed below.

5.2.2    **Examples of energy consumption in aluminium production – primary and secondary materials**

The energy components of aluminium production are given in Table 5.2(a). The energy values are converted to costs per ton of aluminium assuming that 1 ton requires 1 tons of alumina obtained from 4 tons of bauxite. At a conversion efficiency of 23.85% the total requirement is 328 GJ/t. This is the energy cost of producing the crude metal and does not include the energy costs of casting, milling, rolling etc. This estimate is higher than other examples discussed below but demonstrates the balance of energy requirements from mining through to smelting.

| Table 5.2(a) | | |
|---|---|---|
| ITEMS IN ENERGY COST OF ALUMINIUM | | |
| Process | Energy Used per ton Aluminium | |
| | $MJ_e$ | $MJ_t$ |
| Mining (4 tons bauxite) | 42 | 291 |
| Bauxite crushing etc. (4 tons) | 64 | 23 |
| Bayer process (2 tons $Al_2O_3$) | 1,728 | 50,200 |
| Transportation (2 tons) | | 421 |
| Hall-Héroult cell | 54,000 | 43,056 |
| Total | 55,836 $MJ_e$ | + 93,992 $MJ_t$ |
| Equivalent total at 23.85% conversion to electricity | 328 $GJ_t$/tonne | |

Source: P.F. Chapman. The Energy Costs of Producing Copper and Aluminium from Primary Sources, Metals and Materials, 1974.

The energy intensity of deriving ingot from ore provides significant opportunities for energy savings through the reuse of scrap.
R. Barnes [1] has compared the energy intensities in processing
metals from both ore and scrap; the relevant figures are summarised in Table 5.2(b).

This demonstrates the dramatic energy savings to be obtained in
producing ingot from scrap as compared to ore.

| Table 5.2(b) | |
|---|---|
| ENERGY CONSUMED IN PRODUCING ALUMINIUM INGOT (GJ/tonne) | |
| Mining & Benefication | 48.40 |
| Processing | |
| 100% ore | 250.60 |
| 100% scrap | 7.20 |

Source: Robert Barnes.  The Energy Involved in Producing
        Engineering Materials.  Proc. Instn. Mech. Engrs.
        Vol.190, 29/76.

The US Bureau of Mines has estimated the energy requirements for
recycling aluminium scrap from collection centre through preparation, transport and smelting, ending with molten metal, ingots or
other semi-finished forms roughly equivalent to the primary
metal.

The energy requirements are summarised in Table 5.2(c).  It can
be seen that when old cans are recycled for use as new can stock,
the energy requirements are less than for the recycling of mixed
capital scrap.  In terms of energy savings however, both achieve
very significant levels compared to primary - 92% overall and 96%
for cans.

| Table 5.2(c) | | |
|---|---|---|
| ENERGY REQUIREMENTS IN SECONDARY METAL PROCESSING | | |
| Process | Product | GJ/tonne |
| Reverb melting | Ingots (casting alloys) | 15.90 |
| aluminium scrap | Hot metal (casting alloys) | 20.67 |
| Reverb melting | Hot metal (can stock) | 9.20 |
| aluminium cans | | |

Source: US Bureau of Mines 1976.

---

[2]  The Energy Involved in Producing Engineering Materials,
Robert Barnes 1976, Proc. Instn. Mech. Engrs. Vol. 190, 29/76.

Similar energy savings hae been demonstrated by an energy audit undertaken of the British Aluminium Industry [1]. The energy consumed through smelting to hot metal was estimated as 229 GJ/tonne compared to secondary refining of 14.5 GJ/tonne - a saving of 94%.

Boustead and Hancock have also demonstrated the effect of scrap recycling on the production energy for rigid aluminium container sheet (see Figure 5.2(a)). This shows the system energy requirement reducing from 589 GJ/t with total production from primary aluminium, through 360 GJ/t when all in-house scrap is utilised, to 73 GJ/t when all production is from scrap (process and post-consumer). This demonstrates an energy saving of 87% for complete system energy through to the production of sheet.

### 5.2.3  Energy consumption in scrap recovery

There is, of course, energy consumed in the recovery of secondary aluminium which must be taken into account. There are several processes which could be involved:

o   if the scrap is **mechanically sorted** from refuse, along with other materials such as paper, plastic and ferrous metal, the energy involved in this process has to be taken into account and allocated among the materials. The estimate below is based on the energy consumption of a Neuss-type mechanical separating plant (see Appendix C);

o   the aluminium scrap must be **compacted** prior to transporting to the smelter;

o   energy will be consumed in the **transportation** of scrap (100 km by rail has been assumed);

o   at the smelter the scrap is **sorted** and **contaminants** removed. This occurs in a combination of shredders, magnetic separators, sink and swim separators or thermally in 'melting-off' ovens;

o   the scrap is then smelted, usually in rotary kiln;

o   the energy requirements for the above processes have been estimated as follows [2]:

| | |
|---|---|
| Sorting: | 396 MJ/t |
| Compacting: | 72 MJ/t |
| Transport: | 79 MJ/t |
| Mechanical preparation: | 194 MJ/t |
| Total | 741 MJ/t |

[1]  Energy Consumption and Conservation in the Aluminium Industry, Energy Audit Series 1979.

[2]  P. Pautz and H.-J. Pietrozeniuk, Abfall und Energie, Umwellbudesamt. June 1983, Berlin.

62

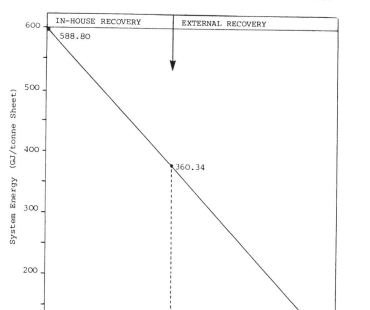

FIGURE 5.2(a):  **THE EFFECT OF SCRAP RECYCLING ON THE PRODUCTION
ENERGY FOR RIGID ALUMINIUM CONTAINER SHEET**

Source:   I. Boustead and G.F. Hancock, Energy and Recycling in
          Metal Systems In Recycling International:  Recovery of
          Energy and Material from Resdiues and Waste.  ed. by
          Karl T. Thome-Kozmiensky, Berlin, 1982.
          ISBN 3-9800462-3-0.

This is a maximum estimate as in certain cases, such as aluminium
cans, there will be no need for any significant final preparation.  Also, there would be some saving in disposal energy,
although this is likely to be small (see Section 5.8.2).

Pautz and Pietrozeniuk have gone on to make a complete system
comparison for producing aluminium from scrap compared to
bauxite, as shown in Table 5.2(d).  Once again very high energy
savings are demonstrated through using secondary materials (94%).
It should also be noted that in comparison with the energy saving
arising from the substitution of secondary aluminium for primary
aluminium, the energy in sorting and preparation is insignifi-
cant.

| Table 5.2(d) | | | |
|---|---|---|---|
| TOTAL SYSTEM ENERGY CONSUMPTION - PRIMARY AND SECONDARY MATERIALS | | | |
| | Electrical Energy (kWh/t) | Thermal Energy (MJ/t) | Total (MJ/t) |
| Aluminium from Bauxite | | | |
| Bauxite production | 500 | 33,400 | 38,530 |
| Electrolysis | 15,000 | 19,530 | 199,770 |
| | 15,500 | 52,930 | 238,300 |
| Aluminium from Scrap 100% | | | |
| Separation and  preparation | 206 | - | 1,950 |
| Smelting | 185 | 3,250 | 5,000 |
| Refining | 190 | 3,360 | 5,160 |
| | 581 | 6,610 | 12,110 |
| Source:  Abfall und Energie (op cit). | | | |

This point is confirmed by a further study undertaken by I. Boustead and G. Hancock [1] on the energy involved in aluminium can recovery by the Alcan scheme in the Dallas-Fort Worth area of Texas. This study however assumed, in line with existing practice in the USA, that the waste cans were transported by car to collection points by can collectors. A round trip distance of 20 miles was assumed. The following stages and associated energies were calculated for the production of 1 tonne of aluminium remelt ingot from can scrap:

|  |  |
|---|---|
| Consumer transport to collection point (20 miles round trip) | 14.52 GJ/tonne |
| Handling at collection point | insignificant |
| Road transport to central warehouse (40 miles round trip) | 0.74 GJ/tonne |
| Magnetic separation and baling | 0.21 GJ/tonne |
| Rail transport to smelter (500 miles) | 0.82 GJ/tonne |
| Smelting (melt to remelt ingot) | 17.30 GJ/tonne |
| | 33.59 |

[1]  I. Boustead and G. Hancock, 'Beverage Containers and Recycling', 1980.

This example illustrates that when collection energy is involved this adds very considerably to the overall energy requirement (43% of total on the above assumptions). But in the case of aluminium, because of the very high energy content in primary production, there are still very high overall energy savings to be obtained.

It is unlikely that the levels of consumer energy in the USA would be the same in Europe. The American system requires driving to collection centres, often out of town, specifically for the purpose of cashing in the cans. In Europe, the systems currently being tested (see Appendix C) are along the lines of the bottle banks where cans are taken to disposal points situated at shopping centres, and are dumped as part of a regular shopping expedition. It is arguable that some of the energy involved in such a trip should be allocated to the can recycling scheme but, even if up to 50% were so allocated, the effect is only marginal in the context of the overall savings available. This is not necessarily the case for other materials.

5.2.4    **Unit energy saving adopted**

All the examples quoted above indicate a high level of savings associated with secondary material producion compared to primary. The range of primary and secondary energy consumptions quoted (excluding Chapman's estimate) were:

> Primary:          219-251 GJ/t
> Secondary:          9-17  GJ/t.

The main savings occur at the smelting stage although, as just noted, if consumer energy is involved this could slightly reduce the savings available. We have adopted a saving of 90% overall for the purposes of calculating gross energy savings. This makes an implicit allowance for consumer energy, for the fact that a small proportion of aluminium substituted will be imported from smelters using hydro-power, and for continued improvement in primary production energy efficiencies.

5.3      **Plastics**

The problems associated with the recovery of mixed plastics from refuse for use as substitutes for virgin polymers have been discussed in Section 3.3 and Appendix B.2. In so far as waste plastics can be reused it is generally for a lower grade use and often in a product which substitutes for another material (e.g. wood or metal). Such alternative uses are still being developed.

It is beyond the Terms of Reference of this study to examine the energy impacts of recovering materials for reuse as substitutes for alternative materials. As mixed plastics offer the only additional recovery opportunity it is their energy value as a fuel or as a by-product from pyrolysis/hydrolysis which are considered in detail (see Section 6.2).

**5.4**     **Paper and Board**

**5.4.1**    **Introduction**

Paper making is energy intensive; up to 15% of the total costs of
production are accounted for by energy costs.  The energy needed
to produce finished paper and board can be separated into the
energy to manufacture pulp and the energy to convert the pulp
into paper and board.  Fuel requirements vary depending upon the
type and quality of product, the type of raw materials used as
well as on the type and size of mill.  Integrated pulp and paper
mills are the least energy intensive because part of the steam
required in paper making can be produced from the waste products
of the pulping process.

The effects on energy consumption of substituting virgin fibre
with waste paper is examined below.  The alternative of using
waste paper to directly recover its calorific value is also
discussed.

**5.4.2**    **Waste paper as a substitute for virgin pulp**
**Examples of energy savings calculations**

There have been a number of analyses undertaken in recent years
to demonstrate the effects of substituting virgin pulp with
secondary fibre.  These provide a range of estimates and generally indicate that recycling provides energy savings.  Selected
examples of recent work are given below.

In Tables 5.4(a) and 5.4(b) the energy consumption for different
mixes of virgin and waste paper is shown for the production of
newsprint and linerboard in an integrated mill.  In both cases
the energy consumption is markedly lower when recycled fibre is
used - 31% and 23% respectively.

An alternative analysis, based on different processes, for
newsprint is shown in Table 5.4(c).  This shows a higher level of
electricity consumption and higher electricity savings but the
overall fuel savings are less (17%) compared to the example in
5.4(a).

Table 5.4(a)

ENERGY CONSUMPTION IN NEWSPRINT MANUFACTURE

| | | V = 1 | V = 0.67 | V = 0 |
|---|---|---|---|---|
| Wood preparation, | S | 3,800 | 2,900 | 1,000 |
| pulping, deinking | kWh | 1,130 | 760 | 380 |
| | | | | |
| Paper making | S | 5,400 | 5,400 | 5,400 |
| | kWh | 180 | 180 | 180 |
| | | | | |
| Effluent treatment | kWh | 14 | 14 | 36 |
| | | | | |
| Miscellaneous | S | 900 | 1,200 | 1,050 |
| | kWh | 14 | 14 | 36 |
| | | | | |
| Total | S | 10,100 | 9,350 | 7,600 |
| | kWh | 1,340 | 980 | 632 |
| | | | | |
| Total in Gigajoules | | 21.0 | - | 14.5 |

Key:   kWh = kilowatt hours per short-ton; S = lbs of steam per
       short-ton; V = virgin fibre per unit total fibre input
Source: 'Secondary versus virgin Fibre Newsprint', in Pulp and
       Paper, Vol.50, no.5, May 1976.  Taken from Waste Paper
       Recycling, OECD 1979.

Table 5.4(b)

ENERGY CONSUMPTION IN LINERBOARD MANUFACTURE

| | | V = 1 | V = 0 |
|---|---|---|---|
| Wood preparation | S | 90 | |
| | kWh | | - |
| | | | |
| Pulping (chemical or | S | 3,201 | 2,000 |
| secondary) | kWh | 95 | 295 |
| | | | |
| Power and steam generation | S | 3,200 | - |
| for chemical pulping | kWh | 145 | - |
| | | | |
| Effluent treatment | kWh | 27 | 27 |
| | | | |
| Board making | S | 8,000 | 8,000 |
| | kWh | 327 | 72 |
| | | | |
| Miscellaneous | S | 1,600 | 2,000 |
| | kWh | 86 | 195 |
| | | | |
| Total | S | 16,000 | 12,000 |
| | kWh | 770 | 590 |
| | | | |
| Total in Gigajoules | | 28.4 | 21.4 |

Key:   As Table 5.4(a).
Source: 'Economics of Recycled Fibre Usage for Linerboard', Pulp
       and Paper, Vol.50, no.4, April 1976.  Taken from Waste
       Paper Recovery, OECD 1979.

| Table 5.4(c) | | |
| --- | --- | --- |
| ENERGY CONSUMPTION IN NEWSPRINT MANUFACTURE | | |
| | V = 1 | V = 0 |
| Energy content in raw material delivered to mill | 2.81 | 2.63 |
| Energy needed to process raw material into newsprint | | |
| i)  Electricity | 2.00 | 0.93 |
| ii) Fuel | 0.81 | 1.13 |
| Total | 5.62 | 4.66 |
| Total in Gigajoules | 20.2 | 16.7 |

Source: Adapted from data in L. Hanserud and O. Olsson; 'Skall vi Bränna upp eller Atervinna Returpapperet', in Teknisk Tidskrift, 2, pp.18-19.

Further analyses demonstrate higher energy savings. Keller [1] has estimated a 64% saving in producing newsprint from recycling fibres and SVA (Netherlands) estimate almost identical savings in producing waste paper pulp compared to mechanical pulp.

A final example of the level of energy savings attainable is provided by a study undertaken in Canada. The percentage savings range from 36% to 63%, depending on the type of paper produced. The absolute values in Table 5.4(d) should be treated with caution as they are based on delivered energy consumption.

**Location of energy savings**

As noted earlier, the extent to which energy savings accrue to the EEC through recycling depends on the extent to which different parts of the paper production chain occur within the EEC. In the examples quoted so far, the energy savings relate to situations in which the entire process from timber felling to paper manufacture occur. In the case of the EEC this is far from being the situation; wood pulp is extensively imported. In such cases the energy tied up in pulp production is not available for saving within the EEC. As can be noted in Tables 5.4(a) and 5.4(b), it is at the pulping stage that the significant energy savings are obtained.

[1]  I.R. Keller, 'How to Establish a Recycled Paper Purchasing Program', Maryland Dept. of Natural Resources, 1980.

| Table 5.4(d) | | | | | |
|---|---|---|---|---|---|
| ENERGY SAVINGS USING RECYCLED PAPER IN PRODUCTION (GJ/tonne) | | | | | |
| | Energy Consumption | | | | |
| Paper<br>Product | Using<br>100%<br>Virgin<br>Pulp | Using<br>100%<br>Waste<br>Paper | Using<br>83%<br>Waste<br>Paper | Using<br>34%<br>Waste<br>Paper | Energy<br>Savings<br>by Using<br>Waste Paper |
| Newsprint | 32.5 | 27.0 | N/A | N/A | 11.8 |
| Printing<br>& Writing | 65.0 | N/A | 37.1 | 55.7 | 9.3-27.9 |
| Tissue &<br>Sanitary | 65.2 | 23.7 | N/A | N/A | 41.5 |
| Corrugated<br>Container<br>Board | 35.2 | 21.3 | N/A | N/A | 13.9 |

Source: Environment Canada, Net Energy Savings from Solid Waste
       Management Options, Ottawa, 1976.
Note:   N/A - not applicable or not available.

FIGURE 5.4:   **ENERGY REQUIREMENTS FOR SELECTED PULP AND PAPER
             PROCESS/PRODUCT COMBINATIONS (NEW MILL BASIS)**

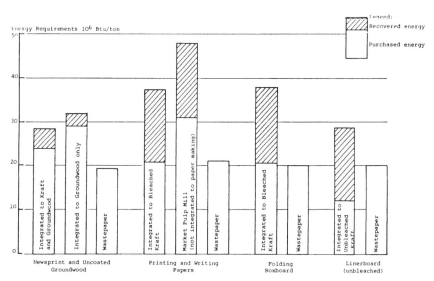

Source:   US Environmental Protection Agency, Pulp and Paper Industry Report, Volume V of Environmental
          Considerations of Selected Energy Conserving Manufacturing Process Options, EPA-600/7-76-034e,
          December 1976.  Taken from Iannuzzi (1977).

## 5.4.3 Unit energy savings adopted

There is some ambiguity in the exact extent of energy savings available through waste paper recycling. Based on the examples reviewed here and elsewhere an average saving of 28% has been taken. This is possibly conservative but is considered prudent given the uncertainties involved; it also allows for the increasing energy requirements of waste paper preparation, such as deinking, as poorer qualities of wastes are brought into the recovery cycle. Also to utilise all the waste paper available would require alternative product uses which at present are uncertain. These may or may not provide similar energy savings in relation to the products they replace.

## 5.5 Glass

### 5.5.1 Introduction

It has long been established that the energy required to melt cullet is less than the energy required to produce glass from virgin raw materials. Conventionally glass furnaces have been operated with a furnace feed typically containing 20% cullet, most of which comes from in-house sources. As the amount of cullet available from bottle banks and other sources grows, there will be opportunities for increasing the proportion of cullet in glass manufacture. The energy implications of this are examined below.

### 5.5.2 Energy in glass container production

Figure 5.5(a) shows the different stages involved in the production of glass containers and the recycling of cullet. The analysis here concentrates on glass containers as it is the recovery of container cullet which offers by far the greatest potential for further material recovery. The unit energy involved in each of the main stages is shown in Table 5.5(a). The following points should be noted:

o  **Recovery of foreign cullet:** this refers to cullet provided be local suppliers under long standing agreements;

o  **Glass factory operations:** the energy associated with the factory operations is considered in two parts - those which are affected by the proportion of cullet in the furnace feed and those which are independent of cullet concentration;

o  **Cullet processing:** before collected cullet is passed to the glass factory for use in the furnace it is treated to remove potentially damaging impurities, particularly aluminium, ferrous metal, ceramics and other large contaminants;

o   **Industrial cullet collection:** the unit energy estimate includes allowance for the provision of the collection bin and transport of cullet to the central storage area;

o   **Conventional bottle bank:** included here is allowance for collection from the bottle bank skips and transport to the central storage area, the energy to produce the collection skip and the trunking of cullet to the processing plant. The latter assumes transportation by road to the processing plant using 20 tonne payload vehicles. The distance is based on UK experience where the average distance is 95 miles.

In order to show the effects of different methods of cullet collection on energy requirements alternative collection schemes are shown;

o   **Consumer Energy:** under the assumption that the consumer makes no special journeys to the collection point zero energy has been attributed to this operation. However, if the purpose of a consumer is to visit the bottle bank and to do something else, such as shopping, then a proportion of the energy associated with his journey could reasonably be allocated to the bottle bank visit. Based on data available concerning the habits of consumers visiting bottle banks (mode of travel, distance travelled, quantity of cullet transported, etc.), we have estimated that on a dual purpose trip the consumer transport energy would be 0.5 and 1.5 GJ/t. **Thus the assumptions concerning whether consumer energy should be included, and if so, how much, are highly critical to the calculation of the energy intensity of glass recycling.**

o   **Disposal:** the energy included here is for disposal only. No allowance is made for refuse collection energy on the grounds that it is independent of the mass of material handled and would be incurred anyhow. Obviously there could come a point where the separate recovery of glass and other materials results in the need for a lower level of refuse collection facilities, thereby producing some energy savings (see Section 5.7.1).

| Table 5.5(a) | |
|---|---|
| GROSS ENERGY REQUIREMENTS FOR THE UNIT OPERATIONS IN GLASS CONTAINER PRODUCTION | |
| Operation | Gross Energy in MJ/kg Output from Unit Operation |
| 1. Production of raw materials | 3.290 |
| 2. Delivery of raw materials | 0.174 |
| 3. Provision of foreign cullet | 1.01 |
| 4a. Glass factory (cullet dependent)* | $8.438 - 0.0236P + 869/T$ |
| 4b. Glass factory (cullet independent) | 3.431 |
| 5. Delivery of processed cullet | 0.049 |
| 6. Cullet processing | 0.097 |
| 7. Collection of industrial cullet | 0.247 |
| 8. Conventional bottle bank | 0.297 |
| 9. Circuit collection with steel bins | 0.515 |
| 10. Bottle bank with GRP bins | 0.306 |
| 11. Circuit collection with GRP bins | 0.356 |
| 12. Consumer energy | 0 |
| 13. Disposal | 0.087 |

Note:   *$P$ = % cullet in furnace feed
          $T$ = throughput of furnace in tonne/day

Source: I. Boustead and G.F. Hancock. 1982. 'Energy Savings
          Through Glass Recycling'; Undertaken for the Glass
          Manufacturers Association.

**FIGURE 5.5(a):   SCHEMATIC FLOW DIAGRAM OF OPERATIONS IN GLASS
                   CONTAINER PRODUCTION WITH CULLET RECYCLING**

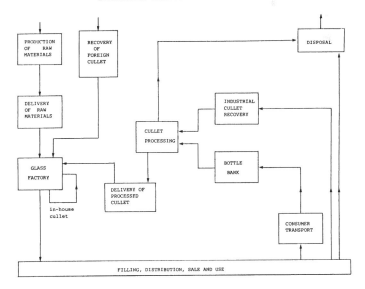

5.5.3    **Energy savings available through cullet substitution**

Based on the above values the overall energy savings of intro-
ducing different percentages of cullet into the furnace feed can
be calculated.  These are shown in Figure 5.5(b).  At present,
20% of cullet is typically used in production (mainly from in-
house sources) and should be used as the base on which to
estimate the effects of further cullet usage.  As can be seen
from the table, the energy savings range from 3.5% when the
cullet concentration is increased to 30% and to 25.5% for 100%
cullet feed.

It should be emphasised that this excludes any allowance for
consumer energy in transporting glass to the bottle banks.  At
the 30% cullet level, an allocation of approximately 7% of this
energy would negate the energy saving.

FIGURE 5.5(b):    **EFFECT OF DIFFERENT PROPORTIONS OF CULLET IN
FURNACE FEED ON TOTAL SYSTEM ENERGY TO PRODUCE
1 kg OF WHITE GLASS AS CONTAINERS IN UK**

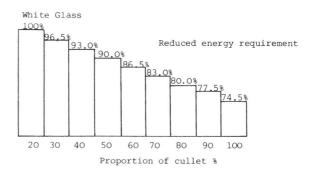

Source:   Bousted and Hancock (op cit).

As noted, the energy savings summarised in Figure 5.5(b) reflect
UK circumstances.  Other studies concerned with estimating energy
savings from cullet recycling have been conducted in Italy and
Germany.

In 1981 research [1] was undertaken in Italy to determine the
amount of energy saved when the cullet proportion was increased
in the production of green and flint glass containers.  Both
furnaces were end-fired by fuel oil.  Common commercial cullet
was used to produce green glass and factory culet for the flint.

[1]   Research carried out during 1981 under an agreement between
Assovetro (Italian Glass Manufacturers Association) and CNR
(Italian National Research Centre).

The energy consumption observed at different levels of cullet usage are shown in Figure 5.5(c). A base figure of energy consumption at 30% cullet utilisation was adopted. For the green glass the average reduction in energy for each 10% increase in cullet was 2.3% and 3.2% for flint glass. As can be noted the relationship is not linear; the energy savings decline as the proportion of cullet increases.

A direct comparison cannot be made between the English and Italian studies as no account of raw material production and transportation energy savings were taken into account in the Italian study. This would tend to increase the overall energy savings assuming (as in the case of the UK study) no allowance is made for consumer energy in transporting cullet to bottle banks.

**FIGURE 5.5(c):**  **EFFECT OF DIFFERENT PROPORTIONS OF CULLET IN FURNACE FEED ON ENERGY TO PRODUCE GREEN GLASS AND FLINT IN ITALY**

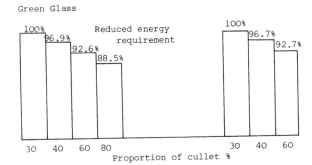

Source:  Assovetro & CNR (op cit).

## 5.5.4  Unit energy savings adopted

There is a  broad measure of agreement that the energy savings available from increasing the proportion of cullet to 100% in glass production would be around 25%. This is the unit energy savings figure which has been adopted. It clearly represents an absolute maximum; the achievement of half this target (50% cullet usage) would represent a very sharp change in manufacturing practise.

## 5.5.5  Reusable containers

An alternative to cullet recovery is to recycle the container for reuse. This, of course, was general practice before the advent of non-returnable bottles. Deposit bottles were returned empty to stores who in turn sent them for refilling. However, during the 1970's non-returnable containers for certain beverages became strongly established. A number of factors contributed to this:

o         distribution distances increased with the construction of large-scale bottling and canning plants;

o         labour costs increased relative to capital and material costs;

o         modern consumers preferred convenience of non-returnables.

There is considerable debate on the relative cost and environmental benefits of returnables compared to non-returnables, but it is clear that trippage rate (the number of times a bottle is used) is crucial to the argument. Each individual returnable bottle absorbs more raw materials and energy than its non-returnable, lighter counterpart and also imposes greater retailing and distribution costs (in the form of storage space, etc.). The more times a bottle is returned, however, the more these extra initial costs are offset.

Similarly with energy. As a refillable bottle is reused, the energy associated with its manufacture remains constant as do energies associated with final disposal, which will occur once in its life. In comparison for each sale via a non-returnable bottle there will be the energy associated with its manufacture as well as disposal energy. Returnables generally have a higher weight of glass per litre capacity and consequently a higher initial gross energy requirement (GER) per litre. We have calculated this at 42.85 MJ/l of capacity including disposal. The lighter non-returnable bottle has a GER of 18.132 MJ/l including disposal for an equivalent size bottle (250 ml). The energy required for transport to return the returnable bottle to filler and wash is approximately 1.34 MJ/l. Figure 5.5(d) shows that the higher initial GER of a returnable bottle and the washing energy is offset after a trippage of 2.6 compared to a non-returnable bottle.

**FIGURE 5.5(d): COMPARISON OF GROSS ENERGY REQUIREMENTS FOR 250 ml RETURNABLE AND NON-RETURNABLE BOTTLES**

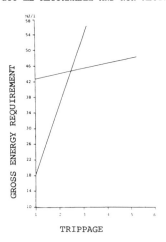

TRIPPAGE

A study [1] undertaken by the Waste Management Advisory Council for the UK Government in 1981 indicated that if all non-returnable bottles were replaced by an all-returnable system, energy savings of 11,600 TJ or 21% would be achieved. This was based on conservative assumptions concerning trippage - 10 for on-premise consumption and 4 for off-premise consumption; higher trippage rates (which are probable) would give even larger energy savings.

The possibility of European legislation to promote returnable drinks containers is now under consideration. A draft directive has been referred to the European Parliament. If this is implemented and supported by national governments significant energy savings could be expected.

## 5.6    Rubber

### 5.6.1    Materials produced

Rubber consumption in the Member States is made up of both natural and synthetic rubber with extensive blending of rubbers in compounds. An example of the extensive compound formulation that occurs is given in Table 5.6(a) which shows the average composition of tyre tread compound.

Following production of natural rubber from latex which occurs largely outside the European Community energy is required for the:

o        production of synthetic rubbers from crude oil;

o        compounding of rubber materials;

o        processing of rubber compound to products.

| Table 5.6(a) | |
| --- | --- |
| COMPOSITION OF TYRE TREAD COMPOUND | |
| | % by weight |
| Natural rubber or SBR (or both) | 50 |
| Carbon black | 30 |
| Process oils | 15 |
| Zinc oxide | 2 |
| Others, as sulphur, accelerators, anti-oxidants etc. | 3 |
| | 100 |

[1]   Study of Returnable and Non-returnable Containers.   HMSO, 1981.

### 5.6.2    Primary process energy

The gross energy requirements for the production of primary
natural and synthetic rubbers are quite different, essentially
due to the considerable feedstock energy content of synthetic
rubber, normally about 42%.  This makes the analysis of rubber
product gross energy requirements difficult in so far as there
are different proportions of natural and synthetic rubber used,
e.g. car inner tubes are virtually all synthetic rubber.
Conversely latex rubber is extensively used in rubber mouldings,
although synthetic latex is taking an increasing share of this
market too.   Trends are, however, difficult to identify due to
the fluctuations in market price of natural rubber and
resubstitution in rubber products.

As discussed in Section B5.2 the major energy savings with rubber
are likely to occur in the retreading of tyres, it is necessary
therefore to consider the primary and remoulding energies
involved.

#### i)      Primary raw materials production

Various estimates of virgin raw rubber gross energy requirements.
The NARI in America [1] has calculated raw rubber energy process
requirements of 36.9 GJ/t.  Boustead and Hancock [2] have further
estimated a requirement of 33.0 GJ/t, which would appear to be in
general agreement.  Boustead and Hancock also calculated energy
requirements for styrene-butadiene rubber (SBR) co-polymer
production on the basis of a total cumulative energy including
feedstock of 16.9 GJ/t of which 10.34 GJ/t was provided by
electricity at an efficiency of 26.5% and the remainder was
provided by oil at 87.5% efficiency at 130 GJ/t.  Further
estimates provided indicate a slightly higher value on the basis
of industrial usage at 145 GJ/t.  On the basis of overall energy
consumption of natural rubber and SBR rubbers, SBR comprising
about 40% of the total synthetic consumption, on the proportions
of 32% and 68% in the EEC leads to calculation of a primary
rubber compound gross energy requirement of approximately 99
GJ/t.

#### ii)     Primary tyre production

In an analysis of primary tyre and remoulding energy requirements
it is necessary to consider the energy content of all materials
included and process energies.

Many of the processes are common to both new and remoulded tyres.
Compounds are mixed, the tyre is built up on a base of textile,
steel or glass and is vulcanised in a mould to impart the
required shape and tread pattern.  Vulcanising brings about the
chemical changes in the components to develop the required

---

[1]  National Association of Recycling Industries Inc., NY Energy
Generation through Recycling, 1980.

[2]  I. Boustead and G.F. Hancock.  Handbook of Industrial Energy
Analysis, John Wiley & Sons, 1979.

strength in the product. In remoulding, carefully chosen used tyre casings are buffed to give the correct profile, then subjected to the same building up and vulcanising processes as new tyres. In both instances the processing is of the batch type, but the manufacture of new tyres is more automated, from the weighing-out of ingredients prior to mixing right through to the warehousing of products.

The Rubber and Plastics Research Association [1] carried out an extensive review of the potential for reclaiming waste tyres. In their analysis, which was subsequently used in the CREST (DG XII) Report on Rubber Waste Recovery in 1979, the energy content of an 'average tyre' was between 0.90 and 1.07 GJ/tyre. Taking the energy efficiency of providing crude oil and process energies of production the potential energy content is between 1.0-1.19 GJ/tyre. This is equivalent to 303.3-360.6 GJ/t of tyres produced.

Further estimates have been supplied for this study by ANRED from an analysis of the potential for waste tyres recycling. These provide gross energy requirement for the 'average' tyre consumed in France of between 266.7-290.9 GJ/t of tyres produced.

Truck tyre production is less automated than car tyre production and generally not carried out by similarly efficient production machinery. Boustead and Hancock (op cit) have calculated a gross energy requirement for a tyre of 39 kg [2] for use on vehicles of payload exceeding 5 tonnes. Radial and cross-ply tyres differed according to their rubber content, radials now comprise the greater share of the commercial vehicle market. The gross energy requirement was calculated on the basis of 100 MJ/kg for all polymers, 50 MJ/kg for steel and process energy requirements of 200 MJ/kg of rubber compound moulded. This equates to a gross energy requirement of about 304 GJ/t of tyres produced.

These values are based on the current compounding technology significant savings could be brought about by:

(a)     substituting natural rubber for synthetic rubber;

(b)     substituting other less energy-intensive materials for rubber in tyres;

(c)     using less rubber per tyre (smaller, lighter);

(d)     increasing tyre durability.

---

[1] A Study of the Reclamation and Reuse of Waste Tyres. RAPRA (UK) 1976.

[2] Figure modified according to International Rubber Study Group data for current commerical vehicle tyre weight.

78

The potential for reduction of rubber waste through these options varies considerably. Option (a) has a potential of reducing the use of synthetic rubber, replacing it with the renewable resource of natural rubber, by 35% beyond current usage patterns. Natural rubber tyres use about 40% less energy, even if the potential fuel value of natural rubber is included. Options (b) and (c) probably have no near-term or immediate term potential. Option (d), which embodies the 160,000 km (100,000 mile) tyre concept, could cut energy use for tyre production by up to 75%.

This substitution of natural rubber (NR), which is a low-energy renewable resource, for synthetic rubber (SR) is a proven technology; NR is actually superior to synthetic rubber in most instances. Currently, about 25% of all rubber used in tyre manufacture is NR, chiefly for heavy duty off-road vehicles. The Malaysian Rubber Bureau, principal research and public relations arm of the NR Industry, claims that NR could comprise up to 60% of all applications without significant technological problems [1]. While SR is generally used for tread material, there are no major technical problems with NR-SR substitution.

5.6.3    **Secondary process energy**

   i)    **Reclaim production**

NARI (op cit) have calculated natural rubber secondary processing energy of approximately 11 GJ/t. Of this about 1.8 GJ/t would be required in the mechanical preparation of reclaim, the remainder used to recombine the product following blending. RAPRA (op cit) put forward a total figure for energy consumption of reclaim production of 15 GJ/t according to American sources.

The process requirement of SBR under secondary processing is about 75 GJ/t not including conservation of feedstock.

   ii)   **Retread production**

The energy required to retread tyres has been widely reported. RAPRA (op cit) considered that between 0.38 and 0.43 GJ/tyre were required for retreading including buffing off of waste rubber, this is equivalent to about 115-130 GJ/t including transport by road for 100 km there and back. This is equivalent to an energy saving of 63%.

Analysis of the French retreading industry led ANRED to conclude that the energy requirement for retreading was between 89-115 GJ/t of tyres produced. This provides a saving of about 175 GJ/t of tyres retreaded which is equivalent to 63%

As with primary production retreading of lorry tyres, which occurs to a higher proportion than car tyres, is more energy-intensive. Boustead and Hancock have calculated a value of 212 GJ/t, including transport energy, representing an energy saving of 30%.

[1]   Private communication.  International Rubber Study Group.

5.6.4    **Unit energy savings**

The following unit energy savings have therefore been adopted for direct material substitution:

| | |
|---|---|
| Natural rubber | 66% |
| Synthetic rubber | 35% |
| Overall rubber consumption in EEC | 45% |
| (at 68% synthetic, 32% natural) | |
| Retreading tyres: | |
|   Car | 63% |
|   Commercial vehicles | 30% |

These are equivalent to 44 GJ/t of normal consumption pattern compound recycled in the EEC, 184 GJ/t of car tyres retreaded and 92 GJ/t of commercial vehicle tyres retreaded.

5.7    **Wood**

5.7.1    **Material production**

The system of wood production as timber is considered in Appendix B.6.

The process is highly complicated by the return of waste materials and production of secondary goods, and by the use of waste materials in provision of energy in the sawmilling industry, principally for drying. In addition there are large variations in the methodology of processing and in particular timber drying. These variations arise from:

o        artificial and natural drying of timber;

o        cold climate increasing energy consumption in drying;

o        the degree of automation;

o        the degree of materials transport and processing,

The energy consumption in timber production is considered below.

5.7.2    **Primary production**

**Wood procurement**

Ignoring the energy involved in planting and maintaining forestry and renewable solar inputs, the energy consumption of procuring wood from the forest is between 100 MJ/m$^3$ for softwood logs and 122 MJ/m$^3$ for pulpwood. Sixty percent of this energy is consumed in long-distance transportation, 27% in forest transportation, 13% in wood cutting. Production of sawlogs is more energy-intensive due to the greater constraints on quality and processing and hence may absorb about 80% of total wood procurement energy in the EEC. Veneer and fibre/particle board absorbs 16% and 3% respectively of total energy expended in wood procurement.

**Sawmill industry**

The energy consumption per unit production has its maximum in drying within the sawmill sector. Total sawmill energy consumption for softwood production amounts to 1.4 GJ/m$^3$, of which 17% is required for size reduction, sawing and planing and the remainder for drying. The source of this energy can be from wood process wastes and levels of energy provision vary from 35% in France and Ireland to between 60 and 84% in Scandinavia. This is heavily dependent on the adoption of integrated facilities for wood processing and product manufacture which are highly automated and process wood to finished products in one unit with heavy utilisation of waste materials generated for energy.

**Total process energy**

The total process energy for wood processing to primary lumber is, therefore, approximately 1.5 GJ/m$^3$ or 2.14 GJ/t at an assumed wood density of 0.7 T/m$^3$.

5.7.3    **Secondary production**

Secondary materials produced from wood waste are not directly substitutable for primary lumber. Secondary materials are principally wood/resin composite products which are used to produce boards and shapes as opposed to the sectional nature of most lumber products. The exception arises in the production of plank (narrow board section) from both lumber and wood-board composites.

The energy required for secondary materials production, assuming that additional energy will be required in the production of fine fibre material for fibreboard and chip, which are directly processed to particle board results in the following energy consumption pattern:

o    Fibreboard          9.98 GJ/t

o    Particle board      3.40 GJ/m$^3$ = 3.78 GJ/t*

o    Plywood/lamin board 6.95 GJ/m$^3$ = 7.72 GJ/t*

*Assumes density of 0.9 t/m$^3$ for composite boards.

The main portion of the energy is heat, comprising 77% of fibreboard production and 85% of plywood.

This reveals that the energy consumption of secondary materials production as opposed to primary production is more intensive. The main reasons for this are:

o    thermal energy requirements;

o    size reduction energy requirements.

5.7.4       **Unit energy savings**

**No unit energy savings are possible for wood processed as a
secondary resource as opposed to production from primary
materials.** Unlike the other materials studied only wood
secondary processing is more energy-intensive than primary
production. The production from primary sources requires between
21 and 63% of the energy required in secondary processing.

Timber reuse direct will save the inherent energy in its
production but transport energy must be taken into account.  If
we consider that primary energy consumption is around 2.14 GJ/t
then the materials can be collected and transported but not
**processed in any way.** Assuming average forest industry
efficiencies for road transport at 0.75 tkm/MJ, net transport
distances of 1,600 km by road for 30 tonne loads could be
accomplished with positive energy returns.

5.8        **Energy in Refuse Collection and Disposal**

It can be expected that the energy associated with refuse
collection and disposal will decrease as increased material
reclamation takes place. The significance of  this saving will
depend largely on the extent of reclamation. The energy
associated with refuse collection and disposal are reviewed
below.

5.8.1      **Refuse collection**

There are two energy components associated with refuse
collection:  that due to the collection vehicles and that arising
from the use of plastic or paper sacks. An estimate of the
energy associated with these two elements, based on UK
experience, has been made by I. Boustead and G. Hancock [1] as
shown in Table 5.8(a).

This indicates that the total average energy to collect 1 tonne
of domestic refuse is 598 MJ. What is interesting to note is
that the energy associated with the provision of plastic sacks is
far more significant than collection vehicle energy; 46% of the
gross energy requirement arises from the use of these sacks.

Collection service energy savings could arise in recovery schemes
where the consumer is required to transport the wastes to a
collection point. Energy savings could be expected when wastes
withdrawn from the refuse stream reach a certain level; the point
at which this occurs will depend on the type of scheme operated.
Local authorities are required to provide a regular collection
service to premises within their area, irrespective of the amount
of refuse generated. A relatively small decrease in the mass of
refuse collected would not reduce the number of sacks and would
only affect very marginally the collection energy. At a certain

[1]   I. Boustead and G.F. Hancock, 'Beverage Containers and
Recycling', 1980.

Table 5.8(a)

ENERGY ASSOCIATED WITH THE COLLECTION OF 1 TONNE OF DOMESTIC REFUSE IN GREAT BRITAIN

| | Electricity (MJ) | | Oil Fuels (MJ) | | | Other Fuels (MJ) | | | Total Energy (MJ) |
|---|---|---|---|---|---|---|---|---|---|
| | Prod'n and Delivery Energy | Direct Energy Use | Prod'n and Delivery Energy | Direct Energy Use | Feedstock Energy | Prod'n and Delivery Energy | Direct Energy Use | Feedstock Energy | |
| Fuel for collection vehicles | – | – | 26.35 | 159.31 | – | – | – | – | 185.66 |
| Ancillary services for collection vehicles | 8.48 | 2.69 | 8.88 | 46.09 | 0.5 | 3.49 | 21.25 | – | 91.38 |
| Provision of dustbins | 5.65 | 2.16 | 0.79 | 3.8 | – | 2.6 | 16.52 | – | 31.52 |
| Provision of plastic sacks | 90.47 | 30.18 | 17.37 | 87.21 | 52.3 | – | – | – | 277.53 |
| Provision of paper sacks | 2.76 | 0.92 | 0.57 | 2.82 | – | 0.01 | 2.38 | 2.56 | 12.02 |
| Total Energy | 197.36 | 35.95 | 53.96 | 299.23 | 52.8 | 6.1 | 40.15 | 2.56 | 598.11 |

Source: I. Boustead & G.F. Hancock, 'The Effect of Cullet Recycling on the Resources Requirements for Glass Container Production', Nov.1982.

level of recovery, however, the local authority should be able to reduce collection frequency or reduce the size of vehicles and thereby save energy. Similarly the number of bags required should reduce.

However, it is possible the recovery scheme might require additional containers for segregating wastes and the energy requirement involved would have to be debited against the recycling scheme.

The average energy consumption does serve to indicate the degree of energy savings which could be available with a significant level of recycling. However, this would only apply where the consumer transport substitutes for refuse collection and when none of the consumer energy is taken into account on grounds that it would be incurred anyhow (for shopping etc.). As already noted, such an approach is questionable and it is unlikely that any significant energy savings can be considered in this context.

5.8.2    **Refuse disposal**

A calculation of the average energy to dispose of 1 tonne of domestic refuse has also been prepared by Boustead and Hancock (op cit). The overall energy required, based on a sample of local authorities in Great Britain, is shown in Table 5.8(b).

Energy requirements obviously vary by type of disposal method adopted. Examples of different unit operations used in refuse disposal are given in Table 5.8(c).

As is to be expected, the energy requirements of incineration are considerably greater than for landfill; pulverisation is slightly higher than landfill.

In the absence of more specific information on the marginal energy requirements of disposal, the average of 135 MJ/tonne provides an indication of the energy savings available when recycled materials reduce to quantities of refuse for disposal.

| Table 5.8(b) | | | |
|---|---|---|---|
| AVERAGE ENERGY REQUIRED TO DISPOSE OF 1 TONNE OF DOMESTIC REFUSE | | | |
| Fuel Type | Production and Delivery Energy (MJ) | Energy Content of Fuel (MJ) | Total Energy (MJ) |
| Electricity | 27.08 | 9.03 | 36.11 |
| Oil fuels | 8.3 | 42.11 | 50.41 |
| Total | 35.38 | 51.14 | 86.52 |

Based on the fuel production efficiencies:
electricity = 25%
oil       = 83%
other fuels = 92%
Source: Boustead & Hancock (op cit).

| Table 5.8(c) | | | | | |
|---|---|---|---|---|---|
| ENERGY REQUIREMENTS PER TONNE OF REFUSE TREATED BY DIFFERENT UNIT OPERATIONS IN REFUSE DISPOSAL | | | | | |
| Type of Operation | Electricity(MJ) | | Oil Fuels(MJ) | | Total Energy (MJ) |
| | Prod'n & Delivery | Energy Content | Prod'n & Delivery | Energy Content | |
| Landfill (a) | - | - | 10.71 | 54.55 | 65.26 |
| (b) | - | - | 22.93 | 116.05 | 138.98 |
| Pulverisa-tion (a) | 116.31 | 38.77 | - | - | 155.08 |
| (b) | 54.22 | 28.07 | - | - | 112.29 |
| (c) | 52.14 | 17.38 | - | - | 69.52 |
| (d) | 65.73 | 21.91 | - | - | 87.64 |
| (e) | 164.16 | 54.72 | - | - | 218.88 |
| Incinera-tion (a) | 621.45 | 207.15 | - | - | 828.60 |
| (b) | 585.86 | 195.29 | - | - | 781.15 |
| (c) | 95.72 | 31.91 | - | - | 127.63 |
| (d) | 716.19 | 238.73 | - | - | 954.72 |
| (e) | 258.75 | 86.25 | - | - | 345.00 |
| (f) | 409.80 | 136.60 | - | - | 546.40 |
| Transfer Station (a) | 72.86 | 24.29 | 18.35 | 92.75 | 208.23 |

Source: I. Boustead & G.F. Hancock, 'Beverage Containers and Recycling', 1980.

## 5.9 Transport Energy

Transport energy is involved at several stages of both the primary and secondary material production cycles, and in the subsequent use of these materials. As secondary materials are assumed to substitute for primary materials the energies involved post-material manufacture would be common to both. Thus, of particular interest is the energy involved in the transportation of raw materials prior to the stage of primary material manufacture compared to the transport energy consumed in collecting materials for recovery and their delivery to the secondary material production plant.

Transport energy can be divided into three main components:

o  the energy content of the fuels consumed directly by the vehicle and the associated fuel production energy;

o  the energy needed to construct and maintain the vehicles;

o  the energy needed to provide facilities for the vehicle to carry out its journey (e.g. roads and railway track).

Typical energies involved with road, rail and sea transport are reviewed below.

## 5.9.1    Road transport

On the basis of the three main energy components listed above the energy proportions for road transport are typically:

| | |
|---|---|
| Fuel | 61% |
| Construction & Maintenance | 32% |
| Route facilities | 7% |

The total energy involved in these activities for vehicles of varying payloads is shown per vehicle km in Table 5.9(a).

On the basis of the figures shown in the table, the energy requirements per tonne/km varies between 12 MJ and 1 MJ. In the case of bulk transport of raw materials the value is likely to be in the range 1-2 MJ which over the likely haulage distances involved represents a very small proportion of total system energy.

Table 5.9(a)

ENERGY REQUIREMENTS PER VEHICLE-KM FOR DIFFERENT SIZE GOODS VEHICLES - MJ/km

| | Fuel Production | Energy Content | Feedstock Energy | Total Energy |
|---|---|---|---|---|
| Rigid vehicles | | | | |
| 1-2 | 2.49 | 9.53 | .01 | 12.03 |
| 4 | 3.12 | 11.92 | .02 | 15.06 |
| 5-8 | 3.61 | 13.79 | .02 | 17.42 |
| 10-12 | 3.90 | 15.26 | .03 | 19.19 |
| 13-20 | 4.24 | 16.18 | .03 | 20.45 |
| | | | | |
| Articulated vehicles | | | | |
| <10 | 4.00 | 15.11 | .03 | 19.14 |
| 11-12 | 4.12 | 15.73 | .03 | 19.88 |
| 13-14 | 4.91 | 18.71 | .03 | 23.65 |
| 15-16 | 4.96 | 18.88 | .03 | 23.87 |

Source: I. Boustead & G.F. Hancock, 'Handbook of Industrial Energy Analysis', 1978.

## 5.9.2    Railway transport

A similar calculation for railway transport produces an overall lower value for freight traffic of 0.8 MJ/tonne/km. Thus the energy element in rail freight is even less significant than for road transport, assuming similar transport distances.

### 5.9.3    Sea transport

The unit values for sea transport are lower still, but the distances travelled are likely to be considerably longer.  The unit energy values range from 0.077 MJ/tonne/km for a 250,000 tonne vessel to 0.233 MJ/tonne/km for a 20,000 tonne vessel. Taking bauxite/alumina as an example, around 60% of production enters intercontinental trade.  An analysis of world trade shows that the average distance travelled by a ton of alumina is about 1,500 miles with an energy requirement of 211 MJ/tonne.  As already noted (Table 5.2(a)) this is completely insignficant in the context of total system energy.  It is also worth noting that this energy occurs outside the EEC and does not necessarily provide an energy benefit to the EEC if replaced by an internally produced secondary material.

### 5.9.4    Conclusion

Bulk transport energies are not a significant factor in the comparison of primary and secondary material energy systems. Consumer transport energy can however be an important factor depending on how this is treated in the analysis.

**6.       THERMAL CONVERSION OF RECOVERED MATERIALS - UNIT ENERGY VALUES**

**6.1      Introduction**

The alternative to the recycling of combustible materials is to
recover their direct energy potential.  It is necessary,
therefore, to consider the potential for thermal conversion of
materials:

o        before recycling; or

o        following sequential recycling.

Thermal conversion is the conversion of inherent energy or
calorific value to heat; it is a finite process whereas recycling
may occur on several occasions.  This section identifies the
energy content of the various commodities which, due to their
chemistry, retain significant calorific values.  The possible
thermal applications of each material is examined separately
followed by their use as part of refuse-derived fuel.

**6.2      Plastics**

As noted earlier, the high levels of purity required for polymer
recycling are such that only small quantities of these materials
can be recovered from mixed waste streams.  Considerably higher
levels of recovery are possible however for the production of
plastic fuel derivates as the sorting system does not have to be
so selective.  Alternatively, the total waste volume can be
**incinerated in bulk** thus removing the need to separate and
classify the wastes.

**6.2.1    Calorific value of plastics**

Plastics contain an element of petroleum feedstock in their
composition and require fossil fuel energy at various stages in
their fabrication.  However, the energy required to process
polymer resins into products is relatively small compared to the
feedstock energy requirements.  Table 6.2(a) provides data on the
relative energy provision of selected plastics products.

Thermal conversion through incineration will not realise the
total feedstock energy of the plastic for three reasons:

o        there will always be production losses, generally 1 to
         10%, in the production of polymer resins; the feedstock
         input must therefore be adjusted to apply to the level
         of true production;

o        a certain fraction of the plastics energy content will
         be utilised in the thermal conversion, e.g. latent heat
         of fusion;

| Table 6.2(a) | | |
|---|---|---|
| ENERGY CONTENT OF SELECTED PLASTICS PRODUCTS | | |
| | Energy Content (%) | |
| End Product | Feedstock | Fabrication |
| Polyvinyl chloride | | |
| - 0.5 gallon container | 85 | 15 |
| High-density polyethylene | | |
| - 1 gallon container | 90 | 10 |
| Low-density polyethylene | | |
| - 1 gallon container | 94 | 6 |
| Polystyrene - meat tray | 83 | 17 |

Source: J. Milgrom, S.R.I. International. 'An Overview of
Plastics Recycling. Paper presented to 'Reclaim
Plastics - Minimise it, Utilise it', Regional Technical
Conference of the Society of Plastics Engineers, Inc.
Oct.14/18, Toronto, Canada.

o        plastics, generally arising in mixed waste streams, will
         be contaminated with other materials and possess a
         moisture content which will reduce the energy
         realisable.

In addition there will be further plastic and feedstock losses on
conversion of the plastics resins.

Table 6.2(b) shows the various aspects of energy recovery from
plastics for the major polymers; thermoplastics comprise 80% of
total EEC consumption. These data show that for polyethylene
(low and high density) of a total gross energy requirement (GER)
of 94-98 GJ/t about 42.00 GJ/t can be realised on combustion of
the polyethylene content of mixed waste. Similarly 19.00 GJ/t of
PVC can be realised on combustion in spite of similar GER to
polyethylene due to the lower feedstock energy and higher
processing requirement.

6.2.2    Direct incineration

Relatively unmixed/clean wastes

Due to the reduction in calorific value of plastics on absorption
or contamination by waste or other materials, in particular wet
organic materials, it is important to realise the energy
potential at the earliest stage possible. In practice this is
difficult to achieve. There are few sources of material that are
sufficiently mixed in polymer content yet arise separately from
other wastes making direct incineration more attractive than
recycling. Possible sources of such categories of material are:

| Table 6.2(b) | | | | |
|---|---|---|---|---|
| ENERGY CONTENT AND CALORIFIC VALUES OF MAJOR PLASTICS (GJ/t) | | | | |
| | GER (1) | Feedstock Energy Content | Calorific Value (HHV) (2) | Calorific Value from Mixed Waste (HHV) (2) |
| **Thermoplastics** | | | | |
| Polyethylene | | | | |
| – Low density | 94.08 | 46.62 | 45.79 | 42.00 |
| – High density | 97.86 | 47.46 | 45.79 | |
| Polyvinyl chloride | 98.83 | 24.20 | 22.69 | 19.00 |
| Polystyrene | 108.46 | 54.36 | 36.79 | 20.00 |
| Polypropylene | 110.16 | 51.20 | 49.14 | 45.00 |
| **Thermosetters** | | | | |
| Polyester (PET) | 153.88 | 46.56 | 45.00 | E 42.00 |

Notes:
(1)    GER - Gross Energy Requirement
(2)    HHV - The HHV of a substance represents the total amount of heat in a unit mass fuel taking into account the negative weight of moisture content of fuel tested.

Sources:
(1)    M. Murat et al, Choice of Methodology of Treatment according to the Physiochemical Characteristics of Industrial Wastes. Recycling International, Ed. K.J. Thomé-Kozmiensky. ISBN 3-9800462-3-0.
(2)    B. Enhorning. The Costs and Value of RDF, Procedures of 2nd Symposium Materials and Energy from Refuse, Antwerp, Belgium, 20-22 Oct.1981.
(3)    J.F. Ingren Housz. An Energy Balance for Plastics Recycling in Household Waste Management in Europe. Bridgewater & Lyon.

o    retail outlets

o    warehousing

o    factories

o    agriculture/horticulture

o    special arisings, e.g. airport wastes.

In our experience other wastes are normally present, e.g. paper, board and wood, and separation would not be desirable or necessary as:

o    the other combustible wastes have an energy value;

o       an energy requirement of between 3 and 15 GJ/t would be
        required to separate the materials mechanically,
        reducing the energy efficiency significantly.

The attraction of direct incineration lies in the availability of
a number of proprietary modular incineration units of both
conventional and starved-air design.  In addition a wide range of
capacities are available from around 250 kg/h to several tonnes
per hour.  Such incinerators can produce both low and high grade
steam for heat or electricity generation for owner needs.

Nominal rates of heat recovery are typically around 50% thermal
efficiency.

### Mixed wastes (municipal wastes)

Direct incineration of mixed waste streams, more notably
municipal wastes, with energy recovery is an accepted technology;
around 15% of municipal wastes are treated in this manner.

In general, the municipal waste incinerator plants with heat
recovery, typically for both heat and power generation, operate
at inputs of greater than 30 tonnes per hour.  At these rates of
input and reuse of waste calorific value (mean 7.5 GJ/t), boiler
efficiencies in the range of 65-75% can be achieved.  It is
assumed that the majority of plants are involved in only the
minimum of waste pretreatment (shearing/ pulverising of bulky
wastes only), but possess electrostatic precipitators or high
efficiency scrubbers to conform to air emission requirements and
operate at availabilities between 75 and 90%.

### 6.2.3   Pyrolysis

Pyrolysis may be an appropriate treatment for a variety of
plastics or plastics-containing wastes and, as with bulk
incineration, only a minimum of waste pretreatment may be
necessary.

Pyrolysis can be adapted to the type of products required via
process changes or, in particular, temperature of degradation and
subsequent product processing, e.g. gas cracking.

In Appendix D.1 the current status of technical developments is
discussed.  Based on this a 55% conversion efficiency has been
adopted.

### 6.2.4   Thermal conversion

The substantial quantities of plastics wastes arising in both
municipal and post-consumer waste streams would probably lend
themselves best to conversion by both direct incineration and
pyrolysis.  A certain fraction of post-consumer wastes of
relatively homogeneous and predictable characteristics, however,
may well lend itself to smaller scale incineration as opposed to
pyrolysis, which requires larger volumes and higher costs.

Table 6.2(c) assesses the energy yield of various thermal conversion options for plastic wastes.

| Table 6.2(c) | | | |
|---|---|---|---|
| ENERGY YIELDS OF THERMAL CONVERSION OPTIONS FOR PLASTICS WASTES | | | |
| Treatment Process/Option | Energy Content (GJ/t) (1) | Efficiency of Conversion (%) | Yield (GJ/t) |
| Bulk incineration | 34.2 | 70 | 23.9 |
| Small-scale incineration | 34.2 | 50 | 17.1 |
| Pyrolysis | 34.2 | 55 | 18.8 |

Notes:

(1)    Energy content calculated on basis of waste composition (W/W)

$$\left.\begin{array}{l}PE)\\PP)\end{array}\right\} - 60\%$$

PS  - 20%

PVC - 15%

REST - 5%

$$= \frac{(60 \times 42.00) + (20 \times 20.00) + (15 \times 19.00) + (5 \times 42.00)}{100}$$

= 34.2 GJ/t

The table shows that bulk incineration is a more efficient means of converting the plastics content of municipal wastes to energy by 5.1 GJ/t over pyrolysis.

In Section 7, in making the calculation of energy savings, the following process combinations have been assumed for post-consumer wastes:

o      bulk conversion

o      50% bulk incineration, 50% small-scale incineration;

o      50% small-scale incineration, 50% pyrolysis;

Again, bulk incineration proves the most energy-efficient option for post-consumer wastes although good returns are available by both mixed scale incineration and pyrolysis which may be a valuable option.

Small-scale incineration of industrially generated waste plastics would be most applicable to industrial wastes.

## 6.3    Paper and Board

Use of paper as a fuel substitute is already widely practiced both as a component of municipal wastes and as part of general factory wastes. The following subsections assess the merit of the use of paper and board as fuel substitutes.

### 6.3.1    Calorific value of paper and board

Paper and board are essentially composed of natural substances and, unlike plastics, have no feedstock energy content.

The calorific value of paper and board is subject to considerable variation due to the potential for absorption of moisture which can considerably reduce the net calorific value. Analysis of municipal waste arisings in the EEC by Gony and Renoux [1] show that dry and wet weight percentage composition of paper and board are 28% and 32% respectively.

This tendency to absorb moisture results in calorific value reductions from as high as 20 GJ/t for dry materials, depending on grade of paper and board down to HHV's [2] of 14 GJ/t and LHV's [2] of 11 GJ/t. It is, therefore, unreasonable to adopt any other value than the LHV for paper and board content of mixed wastes. Industrially generated waste paper and board is less likely to come into contact with moisture from organic or putrescible wastes and possess a higher calorific value if burned on-site.

### 6.3.2    Processes for thermal conversion

#### General

Apart from solid refuse-derived fuels considered separately in Section 6.7, the options for energy recovery by thermal conversion of the paper content of municipal wastes are:

o       bulk incineration

o       pyrolysis.

---

[1]  J.N. Gony and P. Renoux. Assessment of current technology of thermal processes for waste disposal with a particular emphasis on resource recovery. DG12, December 1977.

[2]  The higher heating value (HHV) takes into account the percentage weight of a substance given over to moisture content and the corresponding fall in net calorific value, whereas the lower heating value (LHV) considers both the negative weight of the moisture content and also the energy required for latent heat of vapourisation of the moisture.

**State of technology**

In Section 6.2.1 the present state of technology for bulk
incineration of municipal wastes for plastics wastes was
considered. The same analysis applies to paper and board.

Appendix D.1 considers the specific treatment of plastics by
pyrolysis, although both the Kiener and Babcock systems currently
under development in Germany are purely for treatment of
municipal wastes. As paper waste arisings are entrained in
municipal wastes this is held to be relevant. The other
remaining significant example of pyrolysis processes for
municipal wastes are the Andco-Torrax system which is a **shaft
system** and **fluidised bed-systems** under development in Japan.

**Energy yield**

Energy efficiencies of between 50 and 60% have been adopted for
pyrolysis and 70% has been adopted for bulk incineration of paper
and board.

**6.3.3**     **Thermal conversion**

Adopting the LHV of 11 GJ/t and relative efficiencies of
conversion leads to calculation of the data contained in Table
6.3(a) has been calculated.

| Table 6.3(a) ENERGY YIELDS OF THERMAL CONVERSION OPTIONS FOR PAPER AND BOARD | | |
|---|---|---|
| GJ/t LHV (Calorific Value) | Efficiency of Conversion (%) | GJ/t Yield |
| 11 | 70 | 7.7 |
| 11 | 55 | 6.1 |

Table 6.3(a) shows that the bulk incineration of paper and board
so yields 1.6 GJ/t more than pyrolysis systems.

RDF manufacture and energy recovery is considered separately in
Section 6.7.

## 6.4 Rubber

### 6.4.1 Calorific value of rubber

Natural and synthetic rubbers have different calorific values due
to the significant feedstock energy contained in the latter.
However as there is always a compounding of the two to produce
rubber products it is more realistic to consider overall
calorific values of around 26 GJ/t [1] for the rubber content of
municipal wastes and 31 GJ/t [2] for post-consumer wastes,
principally scrap tyres. The former value is an HHV but, due to
the minimal absorptive capacity of waste rubber, is unlikely to
be reduced significantly on consideration of the heat of vapourisation of moisture retained.

### 6.4.2 Processes for thermal conversion

#### General

The basic methods for thermal conversion of rubber wastes to
energy are:

o       bulk incineration;

o       incineration in relatively smaller units;

o       pyrolysis;

o       use as supplementary fuels.

The latter are generally incineration systems but the heat
utilisation is more direct, a good example being provided by the
use of tyres in cement kilns. These have significant advantages
in that the heat is directly used without conversion losses and
many industries, such as cement works, are well adapted to
prevention of air emissions.

#### Incineration

Both bulk and small-scale incineration of wastes from industrial
premises can be considered on the same basis as plastics. That
is, the technology is well developed and readily available. Bulk
incineration achieves higher conversion efficiencies both due to
efficiencies of scale and technological improvements to the
systems.

Energy yields of 70% for bulk incineration and 55% for small-
scale incineration have been adopted on a similar basis to
plastics and paper and board thermal conversion.

[1]  B. Enhorning.  The Costs and Values of RDF.  Proc. of 2nd.
Symp. Materials and Energy from Refuse. Antwerp, Belgium, 20-22
Oct.1981.

[2]  D. Pautz.  Obtaining Energy from Refuse.  RI International.
Ed. K.J. Thomé-Kozmiensky, 1982.

**Pyrolysis**

There have been similar developments in waste rubber pyrolysis as
for plastics waste. Developments are more widespread
geographically however and a number of major units are under
construction or about to commence commissioning. Key
developments are the processes due to:

o        University of Hamburg, FRG;

o        Tyrolysis Ltd. (Foster Wheeler Power Products), UK

o        Compagnie General de Chauffe (Pyralox), France

o        Fiat/Intenco, Italy

Historically, as with all other thermal conversion plants for
rubber, pyrolysis plants have experienced problems in securing
consistent and adequate supplies of waste rubber, principally
tyres. Developers are confident of the success of this new
generation of rubber pyrolysis plants but due to the commercial
potential of these developments little data are available on
performance. As for plastics wastes, conversion efficiencies of
55% for rubber wastes have been adopted.

**6.4.3        Thermal conversion**

Adopting calorific values as reported of 26 GJ/t and 31 GJ/t for
rubber wastes from municipal and post-consumer wastes
respectively leads to calculation of Table 6.4(a).

| Table 6.4(a) | | | |
| --- | --- | --- | --- |
| ENERGY YIELDS OF THERMAL CONVERSION OPTIONS FOR RUBBER WASTES | | | |
| | GJ/t HHV Calorific Value | % Efficiency of Conversion | GJ/t Yield |
| Municipal waste/pyrolysis | 26 | 55 | 14.3 |
| Municipal waste/incinceration | 26 | 70 | 18.2 |
| Post-cons.waste/small scale incineration | 31 | 50 | 15.5 |
| Post-cons.waste/pyrolysis | 31 | 55 | 17.1 |

Table 6.4(a) shows that bulk incineration would be the most
suitable technology for municipal waste content whereas pyrolysis
of post-consumer wastes would be energetically more favourable.

**6.5**     **Wood**

**6.5.1**     **Introduction**

The thermal conversion of wood waste presents certain problems as follows:

o     variable composition;

o     variable moisture content (dry weight relationships are used by wood processors while furnace and boiler operators consider wet weight moisture contents normally);

o     volatile matters formed (tars can condense on cool furnace sections), bark is lower than timber in volatiles;

o     high ash contents from certain materials, e.g. bark 5%, whereas timber gives only 1%.

Nevertheless, a good deal of wood wastes arisings from forest product industries are relatively uncontaminated and can be burnt directly in solid fuel boilers.

For some of these, it is necessary to employ specially adapted boilers for wood wastes alone which are capable of sequential drying of the wood wastes and recover the maximum amount of energy possible at smaller scales for integrated industrial facilities. Conversion of coal fired boilers to wood wastes can, however, be accomplished economically.

**6.5.2**     **Calorific value**

Due to the highly variable composition of wood wastes and their moisture content it is not entirely realistic to assume a single value. For the purpose of conducting this analysis however an LHV (lower heating value) of 8 GJ/t has been adopted. The equivalent material would potentially have an HHV (higher heating value) of about 16 GJ/t.

**6.5.3**     **Processes for thermal conversion**

**General**

The basic technologies for thermal conversion of wood wastes are:

o     soil fuel boilers;

o     bulk incineration;

o     small-scale incineration;

o     pyrolysis/gasification.

**Solid fuel boilers**

The greatest economic potential probably lies in the direct
combustion of relatively uncontaminated wood wastes from forestry
industries; mostly within the forestry industries themselves
which have considerable energy requirements.

**Bulk incineration**

Bulk incineration on an equivalent basis to that of general
municipal wastes can only be considered for the wood content of
municipal wastes or a fraction of the post-consumer arisings.

**Small-scale incineration**

This is held to be an appropriate technology option for two of
the three categories of wood waste arisings and excludes
municipal waste content of wood wastes where the wastes are
contaminated or mixed with other materials.

A number of proprietary units exist for combustion of wood wastes
operating on either excess or starved-air conditions. It is
essential, however, that the units are capable of accepting
materials of widely varying calorific values (moisture content)
and hence moving grate systems are particularly suitable due to
the ease of control of grate speed and, hence, combustion. The
scale of calorific variation may be in the order of 100 to 300%.

A number of technological developments have occurred, in
particular the K&K Ofenbau AG unit for wood wastes which imposes
strict control on combustion air flows to compensate for
fluctuations in the calorific value.

Typical efficiencies of these specialised units are between 60-
75% and work is currently underway to improve efficiencies
further, principally by reducing the temperature of outflowing
combustion gases or utilising the waste heat to pre-dry the waste
wood. Optimum efficiencies of 85% could be achieved.

**Pyrolysis and gasification**

Pyrolysis and gasification of wood was a widely used technology
40 years ago but is currently undergoing a revival, principally
in developing countries short of petroleum. Gasification is more
widely adopted as the products are more directly useful for
fuelling internal combustion engines or raising steam.

On economic grounds, it is unlikely that pyrolysis gasification
are likely to be at all widely utilised in recovery energy from
wood wastes within the EEC, although certain German companies
have considerable experience in this technology.

It has been assumed that conversion efficiencies by pyrolysis and
gasification are equivalent to about 65%.

**6.5.4    Thermal conversion efficiencies**

Adopting the LHV of 8 GJ/t given in Section 6.5.2, the data
contained in Table 6.5(a) has been calculated.

This indicates that although the options of small-scale
incineration and bulk incineration of wood wastes arising in
post-consumer together with industrial wastes and municipal
wastes respectively are more energy efficient, technological
developments in the field of gasification have reduced the
differences in efficiency between conventional incineration and
pyrolysis/gasification of wood wastes.

| Table 6.5(a) | | | |
|---|---|---|---|
| AVERAGE ENERGY YIELDS OF THERMAL CONVERSION OPTIONS FOR WOOD WASTES | | | |
| | GJ/t LHV Calorific Value | % Efficiency of Conversion | GJ/t Yield |
| Boiler combustion | 8 | 66 | 5.3 |
| Pyrolysis | 8 | 55 | 4.4 |
| Bulk incineration | 8 | 70 | 5.6 |
| Pyrolysis/gasification | 8 | 65 | 5.2 |
| Small-scale incineration | 8 | 70 | 5.6 |

**6.6    Production and Combustion of Refuse-Derived Fuel (RDF)**

**6.6.1    Introduction**

The current status of RDF development has been summarised in
Section 4 and the main types of RDF products are described in
Table 4.3(a).

There are essentially six viable technologies developed and in
production in Europe and the stage of development is now
relatively advanced.  Further refinements to avoid pellet
fragmentation and bring about fluff drying have led to improved
pellet quality and reduced variation.  Although drying increases
process energy requirements still further it is steadily being
adopted by RDF process developers and has therefore been
considered in this analysis.

**6.6.2    Calorific value of RDF and conversion efficiency**

**RDF  products**

Typical calorific values (as LHV) for pelletised RDF are in the
range 11.60 GJ/t to 15.00 GJ/t.  In order to consider the
relative energy contributions to the material due to the combustible components derived from wastes it is necessary to consider
RDF composition.  Table 6.6(a) presents data derived from an
analysis of the Mannesman Veba Umwelttechnik process under
development at Herten in West Germany.

It should be noted that the rubber content of municipal wastes, around 1% W/W, is lost in the mechanical separation process for producing the Eco-Briq RDF at Herten. Thus the relative energetic contributions of the cellulose materials and plastics is 48% and 38% with the deficit coming from the included fines. This assumes calorific values of 42.0, 19.0, 11.0 and 8.0 GJ/t for polyethylene/polypropylene, polyvinyl chloride, paper/board and wood respectively and a mean RDF calorific value of 16.00 GJ/t.

| Table 6.6(a) | | | |
|---|---|---|---|
| RECOVERY OF COMBUSTIBLE MATERIALS IN RDF FRACTION | | | |
| Component | Waste Input (t/a)(1) | Waste Output (t/a) | % Output |
| Cellulose (Paper, Board & Wood) | 105,000 | 94,500 | 90 |
| Plastics (3) | 18,000 | 14,850 | 83 |

Notes:
(1) Household refuse and bulky household/commercial waste like household refuse, i.e. municipal waste 300,000 t/a total.
(2) RDF production at 135,000 t/a, excludes 15,000 t/a of ferrous scrap recovered concurrently.
(3) Rubber is lost to the residuals in the mechanical separation system as it is not a 'light' component of the waste.

The energy necessary to prepare the RDF and energy fraction realised under conventional incineration must also be included in these calculations. The energy consumed in the Eco-Briq process for mechanical separation, drying of fluff and pelletising is equivalent to 1.70 MJ/t of waste input. This is made up from 80 kWh of electrical energy (generated at an efficiency of 37.5%) and 0.9 GJ/t process heat (including process losses). This provides an overall conversion efficiency of 65% with energy conversion efficiencies of 73% for paper and board and 58% for plastics.

### Thermal efficiency of incineration

Thermal efficiencies of boilers operating on coal reduce when burning coal/RDF mixtures. Typically on a 50:50 mix of RDF:coal, thermal efficiency would reduce from 85% to between 75 and 80%.

### Overall thermal efficiencies

The overall thermal efficiency of conversion of the plastic and cellulose content of municipal refuse, on the basis of the experience of the Herten plant in West Germany, can be calculated as below:

|  |  |
|---|---|
| Plastics: | 45% |
| Paper/Board/Wood: | 57% |

#### 6.6.3    Thermal conversion

In the immediate future it would appear unlikely that significant exploitation of waste resources for RDF production outside municipal wastes will occur. We have therefore considered only the production of RDF from the cellulose and plastic content of municipal wastes alone. Table 6.6(b) contains the relevant data.

This table shows that by the production of RDF from the remaining unrecovered quantities of plastics and cellulose wastes would yield an additional 15.4, 6.3 and 4.6 GJ/t of plastics, paper/board and wood respectively.

| Table 6.6(b) | | | |
|---|---|---|---|
| UNIT ENERGY YIELDS OF CELLULOSE AND PLASTIC CONTENT OF MUNICIPAL WASTES BY PREPARATION AND COMBUSTION OF RDF | | | |
| | Efficiency of Conversion % | Calorific Value (GJ/t) | Yield (GJ/t) |
| Plastics | 45 | 34.2 | 15.4 |
| Paper & Board | 57 | 11.0 | 6.3 |
| Wood | 57 | 8.0 | 4.6 |

### 6.7    Comparison of Unit Energy Yields by Thermal Conversion Processes

The relative unit energy yields by the appropriate conventional technologies for thermal conversion of the combustible wastes available are summarised in Table 6.7(a).

A comparison is also made between the yield by preparation and firing of RDF where appropriate to the municipal waste fractions available.

In order to show the range of energy yields of both thermally efficient and less efficient processes values for conventional direct combustion/conversion processes, including incineration, pyrolysis and gasification have been shown.

Table 6.7(a) shows that the most energy efficient option generally is bulk incineration of these combustible waste fractions. Pyrolysis is less efficient than bulk incineration but more efficient than small-scale incineration except for wood wastes. Small-scale incineration, although it may be more feasibile and cost-effective than other treatment methods, possesses the lowest energy conversion efficiency of all options at 50%.

The table also shows that in real terms production of RDF, although it affords greater flexibility than bulk incineration, is a less energy efficient process. Similarly RDF is less efficient than pyrolysis for the cellulose and plastic content of municipal wastes and pyrolysis may also afford an equivalent degree of flexibility.

| Table 6.7(a) | | | |
| :--- | :--- | :--- | :--- |
| COMPARISON OF UNIT ENERGY YIELDS OF THERMAL CONVERSION PROCESSES FOR COMBUSTIBLE WASTES | | | |
| Commodity (1) | Calorific Value (GJ/t) | Efficiency of Conversion (%) | Yield (GJ/t) |
| Plastics | 34.2 | 70 (2) | 23.9 |
| | 34.2 | 55 (3) | 18.8 |
| | 34.2 | 50 (4) | 17.1 |
| | 34.2 | 45 (6) | 15.4 |
| Waste Paper | 11.0 | 70 (2) | 7.7 |
| | 11.0 | 55 (3) | 6.1 |
| | 11.0 | 57 (5) | 6.3 |
| Rubber  MW | 26.0 | 70 (2) | 18.2 |
| | 26.0 | 55 (3) | 14.3 |
| PCW | 31.0 | 55 (3) | 17.1 |
| | 31.0 | 50 (4) | 16.5 |
| Wood | 8.0 | 70 (2) | 5.6 |
| | 8.0 | 65 (3/5) | 5.2 |
| | 8.0 | 70 (4) | 5.6 |
| | 8.0 | 57 (6) | 4.6 |

Notes:
(1)     MW - Municipal waste; PCW - Post-consumer waste
(2)     Bulk incineration
(3)     Pyrolysis
(4)     Small-scale incineration
(5)     Gasification
(6)     Refuse-derived fuel (solids, pelletised)

7.      **OVERALL ENERGY SAVINGS AVAILABLE**

7.1     <u>Introduction</u>

In this section the overall energy savings available through
secondary material recovery are calculated. Based on the
estimates of material availability the energy savings can be
calculated using the energy savings estimates and unit energy
values derived in Sections 5 and 6.

It is necessary however to make certain adjustments to both the
quantities theoretically available for recovery and to the unit
values previously calculated before deriving an estimate of net
energy savings available. These adjustments are described below
followed by estimates of the potential energy saving.

7.2     <u>Potential Quantities of Materials Available for Recycling</u>

An adjustment to the estimates of the physically available levels
of waste materials calculated in Tables 3.8(a) and 3.8(b) is
required to allow for losses which are likely to arise from
dispersion and in their handling and processing. In this
subsection the appropriate adjustments to municipal and post-
consumer waste quantities for material recycling are made; in
Section 7.3 similar adjustments are made to quantities available
for thermal recovery.

7.2.1   **Municipal wastes**

The estimates of the **practical** level of aluminium, paper and
glass recovery attainable from municipal wastes are summarised in
Table 7.2(a).

The assumptions used in making these estimates are given below.
The practical recovery levels of plastics, rubber and wood are
considered in the context of direct energy recovery (where waste
paper is also reconsidered).

**Geographical dispersion:** There will always be a proportion of
waste materials which cannot be recovered due to their geographical dispersion. Irrespective of economic considerations, the
organisational difficulties of organising effective recovery from
widely dispersed waste generators (or from highly congested urban
areas) means that a certain proportion of wastes can be
discounted. To allow for this proportion it has been assumed
that it equates to the percentage of EUR 10 population residing
in rural areas, i.e. 20%.

**Handling and processing:** Losses occur in the handling and
processing of secondary materials prior to the point at which
they substitute for primary materials. The extent of this
obviously varies by type of material and degree of contamination.
Recent estimates (see references at botton of Table 7.2(a)) put
the typical range of losses at 15-20% for aluminium and glass and
35-50% for paper. The mid-point of these ranges has been
adopted.

**Potential quantities available:** After making the above adjustments the following quantities can be considered as representing
the practical amounts still available for recovery:

|  | (000 t) |
|---|---|
| Aluminium: | 590 |
| Waste Paper: | 6,050 |
| Glass: | 4,390 |

It should be emphasised that no allowance has been made for
possible non-cooperation or non-compliance in a recovery scheme.
As has been discussed earlier in the report, further recovery
will rely heavily on source separation and consumer cooperation.
Varying degrees of response have been achieved with existing
schemes. If future cooperation depends purely on voluntary
participation it is very unlikely that the practical level of
recovery will be achieved.

**Potential quantities available adjusted for existing thermal
conversion:** So far no allowance has been made for waste paper
which is currently converted thermally in energy recovery
schemes. It is arguable that, at least in the short term, there
could be considerable reluctance to decrease the calorific value
of the feedstock by recycling paper as a material. This makes a
considerable difference to the quantity of waste paper available.
A final column in therefore shown in Table 7.2(a) assuming none
of the paper currently incinerated as part of heat recovery
schemes is available for recycling.

### 7.2.2    Post-consumer wastes

The same assumptions as used for municipal wastes have been made
adjusting post-consumer waste quantities to derive practically
available levels. These are shown in Table 7.2(b).

### 7.3    Potential Quantities of Materials Available for Thermal Conversion

Compared to material recovery for recycling, higher rates of
recovery for thermal conversion should be achievable as less
constraints operate on material suitability. Factors such as
polymer combinations and origins of plastics and wood materials
become irrelevant in thermal conversion.

Geographical dispersion and rural catchments will probably remain
the greatest constraints to widescale adoption of thermal
processing of wastes although less capital-intensive options do
exist for small volumes of isolated waste.

Separate estimates have been made of the potential availability
of plastics, paper and board, rubber and wood for assessment of
the potential for thermal conversion. Taking a level of 80%
recovery to represent a 20% loss due to geographic dispersion and
rural catchment, Table 7.3(a) can be calculated.

Table 7.2(a)

MUNICIPAL WASTE CATEGORIES STILL AVAILABLE FOR RECOVERY
THROUGH RECYCLING (000 t)

| Material | Physically Available (1) | Physically Available adjusted for Dispersion (2) | Handling & Processing Losses (3) | Potential Available | Potential Available adjusted for existing thermal conversion (4) |
|---|---|---|---|---|---|
| Aluminium | 900 | 720 | 130 | 590 | 590 |
| Waste Paper | 16,100 | 10,500 | 4,450 | 6,050 | 808 |
| Glass | 7,100 | 5,320 | 930 | 4,390 | 4,390 |

Notes:
(1)     Municipal arisings less current material recycling. Taken from Table 3.8(a).
(2)     Assumes 20% of total arisings not available for recovery.
(3)     This represents handling and processing losses of 17.5% for aluminium and glass and 42.5% for paper and board. The latter estimates are based on following references: OECD, Waste Paper Recovery, 1979;
        A. Dornay & W. Franklin, Salvage Markets for Materials in Solid Waste, USEPA, 1972;
        C. Thomas, The Paper Chain, Earth Resources Research, 1972;
        J.A. Van der Kuil, Recovery of Materials by Separate Collection of Domestic Waste Components, Institute of Waste Disposal, 1977.
(4)     Assumes that quantities currently incinerated as part of heat recovery schemes, used for RDF or landfilled for gas recovery, are not available.

Table 7.2(b)

POST-CONSUMER WASTE CATEGORIES STILL AVAILABLE FOR RECOVERY (000 t)

| Material | Physically Available (1) | Physically Available adjusted for Dispersion (2) | Handling and Processing Losses (3) | Potential Available | Potential Available adjusted for existing thermal conversion (4) |
|---|---|---|---|---|---|
| Aluminium | 200 | 80 | 14 | 66 | 66 |
| Rubber | 400 | 260 | 110 | 150 | 92 |

Notes:
(1)     Taken from Table 3.8(b).
(2)     Assumes 20% of total arisings not available for recovery.
(3)     This represents handling and processing losses of 17.5% for aluminium and 42.5% for rubber.
(4)     Assumes that quantity currently incinerated as part of heat recovery schemes, used for RDF or landfilled for gas recovery are not available.

| Table 7.3(a) | | | | | | |
|---|---|---|---|---|---|---|
| COMBUSTIBLE WASTE ARISINGS AVAILABLE FOR THERMAL CONVERSION (million tonnes) | | | | | | |
| | Source (1) | Apparent Arisings (Mt) | Fraction Unrecoverable (2) | Resources Available | Currently Recovered (3) | Potential Availability |
| Plastics | MW | 4.5 | 0.9 | 3.6 | 2.4 | 1.2 |
| | PCW | 7.1 | 1.4 | 5.7 | 1.4 | 4.3 |
| | IW | 0.2 | 0 | 0.2 | 0 | 0.2 |
| Waste Paper | MW | 28.1 | 5.6 | 22.5 | 20.6 | 1.9 |
| Rubber | MW | 1.0 | 0.2 | 0.8 | 0.5 | 0.3 |
| | PCW | 0.7 | 0.11 | 0.6 | 0.4 | 0.2 |
| Wood | MW | 1.5 | 0.3 | 1.2 | 1.0 | 0.2 |
| | PCW | 18.0 | 3.6 | 14.4 | 1.8 | 12.6 |
| | IW | 75.0 | 15.0 | 60.0 | 7.5 | 52.5 |

Notes:
(1)     MW  - Municipal waste
        PCW - Post-comsumer waste
        IW  - Industrial waste
(2)     Calculated at 20% due to geographic dispersion and rural catchment.
(3)     Estimated quantities currently recovered either for energy use or for recycling as a material

Under the assumptions given in Table 7.3(a):

o       5.7 Mt of plastics wastes are available for thermal conversion, the majority arising in post-consumer wastes;

o       1.9 Mt of paper and board wastes are available in municipal waste;

o       0.5 Mt of rubber wastes are available split equally between municipal and post-consumer sources;

o       65.3 Mt of wood wastes are available, predominantly from industrial (forest and wood conversion industries) and post-consumer wastes.

## 7.4     Allowance for Energy Savings which Occur Outside EEC

As noted earlier, to calculate the actual energy savings which accrue to the Community it is necessary to adjust for energy savings arising outside its borders. Thus, for example, energy incurred in the extraction and transportation of source materials outside the Community for subsequent processing within, does not give rise to an energy saving to the Community when substituted by a secondary material. Similarly, when waste material is brought into the Community (e.g. waste paper and scrap metal) for re-processing, the energy involved in collecting and transporting the material outside the Community should not be taken into account. Clearly the more that the processing is undertaken outside the EEC the less likely that energy savings will accrue to it.

## 7.4.1 Adjustment factor

To make allowance for external energy the approach followed has been to calculate energy savings as if all energy were consumed within the EEC and then to apply an adjustment factor to arrive at the net EEC energy saving. The details of the calculation of the adjustment factors are given in Appendix E; the factors are shown in Table 7.4(a).

| Table 7.4(a) | | | |
|---|---|---|---|
| ADJUSTMENT FACTORS TO BE APPLIED TO TOTAL ENERGY SAVINGS TO GIVE SAVINGS ACCRUING TO EEC | | | |
| Commodity | Home production (Mt) | Imports (Mt) | Adjustment Factor – percent to deduct from total energy saving |
| Aluminium | 2.2 | 1.2 | 50.2 |
| Waste Paper | 30.9 | 17.3 | 29.6 |
| Glass | 12.6 | 0.9 | 3.6 |
| Rubber | 2.5 | 1.4 | 28.9 |

The approach followed in calculating the adjustment factors was to:

o  determine the proportions of raw materials, semi-processed goods and finished products (for a given material) which are produced inside and outside the EEC;

o  assume that additionally recovered secondary material substitutes pro rata for internal and external production;

o  weight the resulting internal and external production proportions by the appropriate energy intensities.

## 7.4.2 Unit energy savings available to EEC

The adjustment factors in Table 7.4(a) must be applied to the unit energy savings determined in Section 6. This gives the revised unit savings estimates shown in Table 7.4(b).

| Table 7.4(b) | | | |
|---|---|---|---|
| UNIT ENERGY SAVINGS AVAILABLE TO EEC | | | |
| | Total Energy Saving (GJ/tonne) | Adjustment Factor (percent) to Deduct) | EEC Energy Saving (GJ/tonne) |
| Aluminium | 199 | 50.2 | 97.1 |
| Waste Paper | 7 | 29.5 | 5.0 |
| Glass | 6 | 3.6 | 5.5 |
| Rubber | 44 | 28.9 | 31.3 |

## 7.5     Total Savings Available

### 7.5.1     Introduction

Whether a material should be recovered for use as a secondary
material or energy source depends partly on where and how it
arises. In the case of glass and aluminium the position is
straightforward; the only option is as a replacement material.
Likewise with wood where use as a fuel is the only realistic
option in terms of energy savings.

The choice arises with the other combustible materials -
plastics, paper and rubber. It has already been pointed out that
there are obvious advantages in recycling these for recovery as
materials as the option still remains for ultimate use as a fuel.
In terms of maximising energy savings however this is not
necessarily the preferred choice. It is purely on the basis of
energy savings that the selection of recovery method has been
made here.

**Municipal wastes:** In the case of municipal wastes it has been
assumed that aluminium, glass and paper are recovered as
materials while plastics, rubber and wood are recovered as fuel
products. The recovery of plastics as a fuel instead of as a
material is adopted because at present plastics cannot generally
be recycled to substitute for polymers (only as a replacement
material).

As paper has an alternative use as a fuel, the effect of using it
in this manner as opposed to a material is also shown.

Where materials are recovered for recycling (as materials) the
energy saving assumed is that achieved when recycled back into
its original use (cullet to glass, waste paper to paper and
board, aluminium to aluminium). As observed earlier however it
is by no means certain that this volume of recovered material
could be reabsorbed in this way and new product applications (at
least for glass and paper) will be required. As it is not known
to what extent for what products these would substitute it has
been assumed that the energy savings would be the same as direct
recycling.

**Post-consumer wastes:** The recycling of aluminium and rubber has been assumed from post-consumer wastes. But as rubber can also be used as a source of thermal energy its energy value in this context is also considered.

### 7.5.2    Energy savings through material recycling

In Table 7.5(a) the energy savings available from recycling municipal and post-consumer wastes are shown. The following main points should be noted:

o          the total potential savings available are 123 PJ/a of which 90% comes from municipal wastes. Aluminium is responsible for over half the  total savings;

o          if the paper and rubber which are currently combusted as part of existing energy recovery schemes are excluded, the total savings fall to 95 PJ/a. The contribution from paper reduces to only 4% of total savings under this assumption.

Table 7.5(a)

ADDITIONAL ENERGY SAVINGS THROUGH MATERIAL RECYCLING

| Material | Source (1) | Quantity Mt (2) | Saving Unit (GJ/tonne)(3) | Total Saving (PJ/a) |
|----------|------------|-----------------|---------------------------|---------------------|
| Aluminium | MW | 0.590 | 97.1 | 57.29 |
|  | PCW | 0.066 | 97.1 | 6.41 |
| Waste Paper | MW | 0.808 | 5.0 | 4.04 |
| Glass | MW | 4.390 | 5.5 | 24.14 |
| Rubber | PCW | 0.092 | 31.3 | 2.88 |
| Total |  |  |  | 122.79 |

Notes:
(1)    MW = in municipal waste; PCW = in post-consumer waste
(2)    From Tables 7.2(a) and 7.2(b)
(3)    From Table 7.4(b)

7.5.3    **Energy yields through thermal conversion**

The energy yields resulting from thermal conversion by appropriate conventional technology are shown in Table 7.5(b).

The yield from firing RDF derived from municipal waste fractions is also shown.

In order to show the range of energy yields by different processes, upper and lower values have been shown for conventional direct combustion/ conversion processes.

The following main points should be noted:

o    the total energy yield of the combustible waste fractions of all wastes arising following subtraction of the existing quantities recovered as materials and energy would be between 462.5 PJ/a and 525.9 PJ/a;

o    in real terms, production of RDF, although it affords greater flexibility than bulk incineration, is a less energy efficient process. Similarly RDF is less efficient than pyrolysis of the cellulose and plastic content of municipal wastes and which affords an equivalent degree of flexibility;

o    process wood wastes in the wood manufacturing industry provide over half the thermal potential. After wood, plastics offer the greatest potential, particularly from post-consumer sources;

o    rubber and waste paper represent only 4% of the overall potential.

We have also calculated the energy yield under the assumption that materials which are currently recycled could be redirected for direct thermal recovery (Table 7.5(c)). In practice there would always be a certain volume of materials recovery and certain secondary manufacturing processes, e.g. glass cullet in glass production. This means that the maximum energy realisable from thermal conversion of materials would in practice be lower than those shown in Table 7.5(c).

Redirection of paper and plastics from material recycling in municipal wastes to the production of RDF would increase the net incremental RDF energy yield from 31.4 PJ/a to 126.2 PJ/a.

Table 7.5(b)

ENERGY VALUES OF THERMAL CONVERSION PROCESSES FOR COMBUSTIBLE WASTES

| Commodity (1) | Quantity (Mt) | Energy Yield | | |
|---|---|---|---|---|
| | | Conventional Lower (PJ/a) | Conventional Upper (PJ/a) | Pelletised RDF (PJ/a) |
| **Plastics** | | | | |
| MW | 1.2 | 22.8 (4) | 29.0 (2) | 18.6 |
| PCW | 4.3 | 78.0 (3/4) | 103.0 (2) | - |
| IW | 0.2 | 3.4 (3) | 3.4 (3) | - |
| **Waste Paper** | | | | |
| MW | 1.9 | 11.5 (4) | 14.6 (2) | 11.9 |
| **Rubber** | | | | |
| MW | 0.3 | 4.3 (4) | 5.5 (2) | - (6) |
| PCW | 0.2 | 3.6 (3) | 4.7 (4) | - |
| **Wood** | | | | |
| MW | 0.2 | 0.9 (4) | 1.1 (2) | 0.9 |
| PCW | 12.6 | 65.5 (4/5) | 20.6 (3) | - |
| IW | 52.5 | 273.0 (4/5) | 294.0 (3) | - |
| Total | 73.4 | 462.5 | 525.9 | 31.4 |
| Total excluding IW | 20.7 | 186.1 | 228.5 | 31.4 |

Notes:
(1)  MW - Municipal waste; PCW - Post-consumer waste; IW - Industrial waste.
(2)  Bulk incineration
(3)  Small-scale incineration
(4)  Pyrolysis
(5)  Gasification
(6)  Rubber content of municipal waste not present in RDF produced.

| Table 7.5(c) | | | | |
|---|---|---|---|---|
| ENERGY VALUES OF THERMAL CONVERSION - INCLUDING MATERIALS CURRENTLY RECYCLED | | | | |
| | | Energy Yield | | |
| Commodity (1) | Quantity (Mt)(2) | Conventional Lower (PJ/a) | Conventional Upper (PJ/a) | Pelletised RDF (PJ/a) |
| **Plastics** | | | | |
| MW | 2.2 | 41.4 (4) | 52.7 (2) | 34.1 |
| PCW | 5.0 | 89.8 (3/4) | 119.7 (2) | - |
| IW | 0.2 | 3.4 (3) | 3.4 (3) | - |
| **Paper & Board** | | | | |
| MW | 14.2 | 85.9 (4) | 109.3 (2) | 88.9 |
| **Rubber** | | | | |
| MW | 0.5 | 7.2 (4) | 9.1 (2) | - |
| PCW | 0.5 | 7.8 (3) | 8.5 (4) | - |
| **Wood** | | | | |
| MW | 0.2 | 0.9 (4) | 1.1 (2) | 3.2 |
| PCW | 12.6 | 65.5 (4/5) | 70.6 (3) | - |
| IW | 52.5 | 273.0 (4/5) | 294.0 (3) | - |
| Total | 87.9 | 574.9 | 668.4 | 126.2 |
| Total excluding IW | 35.2 | 288.5 | 371.0 | 126.2 |

Notes:
(1)     MW - Municipal waste; PCW - Post-consumer waste; IW - Industrial waste
(2)     Bulk incineration
(3)     Small-scale incineration
(4)     Pyrolysis
(5)     Gasification

### 7.5.4     Summary

In Table 7.5(d) we summarise the combined energy values of material recycling and thermal recovery.  It has been assumed that:

o        in the case of material recycling the wastes which are currently used in direct energy recovery schemes would not be diverted to recycling;

o        in the case of thermal conversion, the materials which are currently recycled would not be available for direct energy processing;

o   where the option exists for material recycling or
    thermal conversion the route yielding the highest energy
    value would be adopted;

o   in all cases, the process yielding the highest energy
    value is adopted.

The total maximum energy available through additional recovery
for the materials reviewed based on these assumptions is 564 PJ/a.
This is the saving available to the EEC. If the energy savings
arising externally are included this increases to 627 PJ/a.

| Table 7.5(d) | | |
|---|---|---|
| COMBINED ENERGY CONTRIBUTION FROM MATERIAL RECYCLING AND THERMAL CONVERSION | | |
| Material | Energy Value PJ/a | Percent |
| **Aluminium** | | |
| MW | 57.3 | 10.2 |
| PCW | 6.4 | 1.1 |
| **Plastics** | | |
| MW | 29.0 | 5.1 |
| PCW | 103.0 | 18.3 |
| IW | 3.4 | 0.6 |
| **Waste Paper** | | |
| MW | 14.6 | 2.6 |
| **Glass** | | |
| MW | 24.1 | 4.3 |
| **Rubber** | | |
| MW | 5.5 | 1.0 |
| PCW | 4.7 | 0.8 |
| **Wood** | | |
| MW | 1.1 | 0.2 |
| PCW | 20.6 | 3.7 |
| IW | 294.0 | 52.1 |
| Total | 563.7 | 100.0 |

Notes:   MW - Municipal wastes;
         PCW - Post-consumer wastes;
         IW - Industrial wastes

8.        OPTIONS FOR GOVERNMENT ACTION TO PROMOTE FURTHER ENERGY SAVINGS

8.1       Introduction

          Any action to promote energy savings through waste recovery
          cannot be viewed independently of overall policy on waste
          recovery.  Energy savings are only an aspect of this whole
          question and may conflict with other requirements.  For instance,
          high energy savings may be obtainable through recovery of a
          particular material but only at very high cost.  If it is policy
          to recover and save energy **at all costs**, then such recovery could
          in theory be pursued.

          Such a situation would seem to have little by way of supporting
          commercial logic.  However, if wider economic and social costs
          and benefits associated with waste disposal and recovery, as well
          as strategic considerations, are taken into account, certain
          additional net financial costs of further secondary materials
          recovery may be seen as justified.  This point is considered
          further below.

          In this section we summarise the options available to governments
          to promote recovery in general to overcome the obstacles high-
          lighted by the study.  We then concentrate on the action which
          would promote energy saving in particular.  In order to put such
          action in the context of waste disposal policy options, we first
          briefly contrast the financial and social viewpoints which can be
          adopted.

8.2       **Financial and Economic Efficiency Approaches to Material Recovery**
          **Schemes**

8.2.1     **Financial approach**

          The decision by a waste disposal authority on whether to under-
          take recovery for material or fuel will normally be taken in the
          context of its overall waste disposal strategy.  The authority
          could logically be expected to opt for the least cost financial
          disposal route(s) from among the alternatives which would meet
          its statutory waste collection and disposal obligations.  That is
          it would adopt market efficiency criterion and adopt the scheme
          which represented the lowest net cost in financial terms.  In the
          case of recovery schemes this would be after considering
          potential revenue from recovered materials and savings in
          disposal costs.

          Landfill has traditionally provided the least cost disposal
          option but as suitable sites become more difficult to locate
          other methods of disposal/recovery are likely to become more
          attractive.

8.2.2     **Economic/social approach**

          The narrow financial perspective of the above approach can be
          contrasted with an approach based on maximising economic
          efficiency.  This broader (national) perspective considers
          additional benefits such as raw material conservation, energy

savings, reduced levels of imports, any increased employment and possible reduced environmental costs of waste disposal. This approach implies that financial cost considerations alone are not a sufficient basis for evaluating a desirable level of recycling activity. To the extent that national governments adopt this view, it is open to them to influence the level of material recovery through a range of various policy instruments. This includes recycling taxes, subsidies, procurement policies, ordinances and laws.

In practice several countries have adopted measures to encourage resources recovery and in Europe many countries are active in pursuing this objective. Similarly the CEC has a number of declared objectives of direct relevance to the promotion of material recovery:

o       The framework Waste Directive (75/442/EEC), OJ No. L194 25 July 1975 covering:

    a)    the prevention and reduction of waste,
    b)    the recycling and reuse of waste,
    c)    the safe disposal of waste which is not recovered.

o       The directive on Waste Oils (75/439/EEC), July 1975 which seeks to prevent environmental pollution and to ensure that waste oil is recovered either as a fuel or processed for reuse as a lubricant.

o       The recommendation concerning the reuse of Waste Paper and the use of recycled paper (81/972/EEC), OJ No. L355/56, 10 December 1981, covering inter alia, the encouragement of the use of recycled paper and board.

The extent to which additional resource recovery and associated energy savings can be achieved will depend partly on how active national governments are in pursuing this objective. Otherwise, additional energy savings through material recovery will only occur when private net costs dictate. In the absence of government intervention, the existing level of recovery (and energy savings) can be expected to be close to the maximum in existing market conditions.

## 8.3       Possible Action for Increasing Material Recycling

### 8.3.1     Introduction

In this subsection we consider action for the further recovery of waste materials for recycling as secondary materials. In Table 8.3(a) we list the options which are available to overcome existing constraints. These options are all geared to increasing the level of recycling in general - that is, they are not energy-specific. We have not identified any specific action in the area of collecting and processing the materials which would promote further energy savings.

| Table 8.3(a) | |
|---|---|
| **OPTIONS FOR GOVERNMENT ACTION TO ENCOURAGE RESOURCE RECOVERY** | |
| Secondary Materials | Some Options for Action to Overcome Obstacles |
| **Aluminium** | Subsidise development of systems for collecting dispersed scrap from post-consumer waste |
| | Study feasibility of separate collection schemes of household waste and other source separation schemes |
| | Subsidise development of technologies for separating and upgrading aluminium scrap from other materials |
| | Distribute information on the value of non-ferrous scrap to waste producers |
| **Plastics** | Subsidise the development of appropriate technology for segregating thermoplastic polymers in waste |
| | Study separate collection schemes suitable for municipal plastic waste |
| | Encourage development of markets for by-products using mixed plastic waste |
| | Encourage product planning to avoid products that are difficult to recover |
| **Paper** | Enforce procurement policies by public bodies to purchase recycled paper products |
| | Subsidise the introduction of necessary equipment into paper mills so that technically acceptable maximum waste paper content is included in paper products |
| | Support separate collection systems for different qualities of waste paper |
| | Encourage long-term contracts between suppliers and industry |
| | Investigate standards of recycled paper and conduct appropriate market development |
| | Subsidise techniques for treating recycled paper to produce higher quality finishes |
| | Subsidise techniques for the decontamination of waste paper |
| | Distribute information on the suitability of recycled paper for consumers' needs. |

continued/...

| Table 8.3(a) continued | |
|---|---|
| OPTIONS FOR GOVERNMENT ACTION TO ENCOURAGE RESOURCE RECOVERY | |
| Secondary Material | Some Options for Action to Overcome Obstacles |
| **Glass** | Encourage continued introduction of bottle bank schemes |
| | Distribute information to households on the benefits of their participation in these schemes and where to deposit their bottles |
| **Rubber** | Discourage cheap import of reclaim and synthetic rubber |
| | Encourage development of markets for by-products using waste rubber |
| | Encourage voluntary collecting schemes for old tyres where these do not exist |

8.3.2    **Aluminium**

**Additional savings available**

The savings available from the further recycling of aluminium have been estimated as 64 PJ/a.  To reach this level will require that the obstacles be overcome which have restricted the further recovery of these high value wastes in the past.

In the case of **post-consumer wastes** these barriers result mainly from the fact that much of this waste is available in very minute quantities in dispersed locations and that much of it arises in combination with other metals and materials.  The first factor makes collection difficult and transportation expensive.  The second factor adds additional and often substantial costs to the recovery and upgrading of these metals before they are suitable for recycling.  The aluminium which is recovered is extracted from post-consumer durables and automobiles.  Economically viable techniques have been slow to develop for significant levels of recovery to be achieved.

Further aluminium recovery depends on progress being made in separation techniques and in improved collection systems for dispersed scrap.

The levels of aluminium in municipal waste are still small, although likely to grow with increasing adoption of the all-aluminium can.  The source separation schemes being tested for the recovery of cans would seem to offer the best opportunity for increased recycling.

**Possible EEC Action**

We consider there are strong economic arguments for a
continuation of the support for activity currently underway in
the EEC for increasing aluminium and other non-ferrous metal
recovery:

o        support in the form of research grants for the develop-
         ment of suitable separation techniques for aluminium
         waste arisings with other materials and metals;

o        examine schemes for collecting or encouraging voluntary
         delivery of small quantities of dispersed scrap and
         consider subsidising these;

o        encourage the establishment of can recovery schemes.

8.3.3   **Plastics**

Because of the problems involved in identifying and segregating
the polymer types in mixed plastics waste the opportunities for
recycling these wastes to replace primary polymers is extremely
limited.  The main opportunity for reuse is as substitutes for
other products in low grade uses.

Obviously there is merit in recycling plastics for reuse and this
should be encouraged (see Table 8.3(a)) but from the point of
view of this study attention is concentrated on plastics recovery
for thermal applications.

8.3.4   **Waste paper**

**Additional savings available**

The savings available from the further recycling of waste paper
have been estimated as 4 PJ/a, when recovered as a material and
excluding redirected materials from existing recovery by thermal
conversion.

As argued earlier, the attainment of this level (economic
considerations apart) is dependent upon increasing demand for it
- either for reuse as paper or in alternative uses.  The level of
demand is affected by technical and quality considerations that
determine the quantities of waste paper that can be included in
recycled paper products and by fluctuations in international
price levels for pulp and paper, for which waste paper substitutes and competes.

Also the level of demand for paper products containing recycled
paper, especially printing and writing paper, suffers from
discrimination among consumers against its quality; and
techniques for upgrading its quality have not yet been
commercially proven.

120

**Possible EEC Action**

Further EEC action should initially be focused on increasing the level of demand for waste paper. We support the submission by the Commission to the Council for a Recommendation concerning the recovery and reuse of waste paper and board (Draft Council Recommendation, 14th May, 1980; Official Journal No. 135/3).

We believe that if further encouragement is given to paper recycling, priority should be given to the following:

o        introduction of procurement policies by public bodies to purchase recycled paper products, as proposed in the Draft Council Recommendations;

o        guidelines on long-term contracts between suppliers of waste paper and the industry, and recommendations to suppliers (e.g. local authorities) that such contracts are negotiated;

o        investigation of standards of recycled paper, particularly printing and writing paper, with a view to further market development and distribution of information to consumers on the suitability of recycled paper.

If the level of demand for recycled paper products increases, we suggest consideration is given to the following:

o        grants to assist with the introduction of necessary equipment into paper mills so that technically acceptable maximum waste paper content is included in waste paper products;

o        grants to assist with the development of techniques for improving the quality of recycled paper products;

o        grants to assist with techniques for eliminating pernicious contraries in waste paper;

o        grants to assist in the development of alternative product applications for waste paper.

8.3.5    **Glass**

**Additional energy savings available**

The savings available from the further recycling of glass as cullet have been estimated as 24 PJ/a. This is the maximum level and its achievement is dependent upon further public participation in bottle bank schemes and the willingness of local glass plants to receive cullet.

The savings of energy through the reuse of glass containers does, however, offer considerably greater scope for energy savings, the exact level depending on the type of container and trippage rates achieved. To the extent that increased use is made of returnable bottles this will generally provide considerable energy benefits.

**Possible EEC Action**

Possible action falls into two main areas:

o        encourage continued introduction of bottle bank schemes through concerted planning on the part of municipal/ regional authorities and the glass industry;

o        encourage reuse of glass containers wherever feasible; in this respect the EEC's proposed Draft Directive for the standardisation of containers will undoubtedly assist.

However, before any action is taken to encourage further glass recycling/cullet reuse, the Commission should seek to establish through consumer surveys:

o        whether or not special journeys are being made to return bottles under existing bottle bank schemes;

o        to what extent would special journeys become more likely if bottle bank schemes are to be extended.

8.3.6      **Rubber**

**Additional energy savings available**

The level of additional energy savings available from further rubber recycling have been estimated as 3 PJ/a excluding materials redirected from wastes currently recovered by thermal conversion.

The major **obstacle** to additional recovery is the low level of demand for used rubber by the rubber industry; the two main factors that depress this level of demand are competition from foreign imports and the limited markets for rubber products that can be made from rubber crumb and reclaim. The costs of producing these products are relatively high and the processes are energy intensive.

**Possible EEC Action**

Actions that the EEC might consider for encouraging more use of waste rubber are:

o        Discouraging imports of rubber reclaim and synthetic rubber whose prices undercut those charged by producers of reclaim within the EEC, perhaps by import taxes.

o        Encouraging, through research grants, the development of markets for by-products containing waste rubber (such as sports surfaces, rubber soles and road surfaces).

o        Discouraging the landfill and incineration without energy recovery of old tyres.

However, it is not considered that such action can be justified through the energy savings likely to be realised by such action.

**8.4**  **Possible Action for Promoting Thermal Conversion of Combustible Material**

In this subsection we consider action to support the further conversion of wastes to energy. First we summarise some of the institutional considerations which are relevant to this question.

**8.4.1**  **Institutional/Organisational Considerations**

**General**

Waste generators producing wastes in such quantities that energy in the form of fuel, heat or electricity surplus to in-plant requirements is available, face a number of potential problems in marketing that surplus to potential users. For example:

o       there may be no potential user within reasonable distance prepared or able to utilise the energy;

o       there may be only one potential user (e.g. an electricity utility) in a position to take the energy and the absence of competition will affect the negotiation of prices;

o       there may be seasonal variations in the demand for energy, but a constant year-round generation of waste.

We comment briefly on the supply of fuel, heat and electricity in turn.

**Fuel**

The use of wastes of suitable calorific value as fuels either in their raw state or after various levels of processing can be an attractive proposition particularly for in-plant use. Supply of waste-derived fuels to another party can however present a number of problems which need to be overcome:

o       the purchaser may demand that the product consistently meets a certain basic specification (moisture content, calorific value, etc.). This may require additional process equipment and/or quality checks. The latter may prove difficult with heterogeneous waste products;

o       continuity of supply may need to be assured;

o       the purchaser is likely to be using the fuel to supplement coal supplies and the waste derived fuel will be the first source to go if the users' demand for fuel falls.

These problems have led to many waste derived fuel producers becoming their own customer in various ways. For example local authorities supplying pelletised RDF to their own district heating boilers and energy management companies acting as middlemen between waste fuel producers and industrial fuel users.

**Heat**

The supply of surplus heat in the form of hot water or steam to
nearby domestic or industrial users can prove attractive in some
circumstances, most usually in the industrial areas of large
conurbations. Seasonal variations in demand can be problematical
and this option is only feasible if relatively large quantities
of surplus heat are regularly available. The main obstacles are
likely to be technical and economic rather than institutional but
the physical establishment of the networks, acquiring wayleaves
over property and planning approvals, will present organisational
demands.

**Electricity**

From an institutional point of view the generation and sale of
surplus electricity often faces one major difficulty, there being
only one potential purchaser in the form of the public
electricity utility. It will generally not be possible for a
company to supply electricity to other industrial or domestic
users. Typically, and certainly at present in Europe, there is
over-capacity in electricity generation and the utilities are not
therefore seeking alternative sources of supply. Utilities may
be interested in accepting feeds to the grid networks to help
meet peak demands, but even then the contribution from an
individual industrial generator may not be sufficiently significant to interest them.

In those cases where utilities are interested in buying
electricity the payments offered are generally below the
utilities' avoided cost of generation. Conversely, the rates
charged to industry for electricity purchased from the utility
will normally be at standard commercial rates.

These problems will persist so long as there is over-capacity in
the electricity utility systems and there is no obligation upon
them to pay the avoided cost of generation for electricity
supplies bought in from industrial generators.

**8.4.2**    **EEC action in respect of thermal conversion**

**Direct boiler combustion:** The principal opportunity here lies in
increased combustion of uncontaminated wood wastes. Constraints
to use are partly the limits of degree of juxtaposition of
consumers and producers of wood wastes (although this often
neglects the significant potential that lies within the forestry
industry itself); but also inertia or shortage of capital on the
part of consumers to convert their boilers to solid fuel/wood
waste systems. When existing boilers are due to be replaced in
these industries the opportunity is presented for conversion to
wood wastes at economic costs. The ability of sawmills to
generate and sell surplus electricity is constrained by the same
institutional barrier noted above.

Member States, with possible encouragement for the EEC, could encourage the conversion to wood waste boilers by carrying out surveys of steam raising energy consumers in forested areas and making the results available to wood waste producers. Such a survey has already been conducted in the UK.

**Bulk incineration**: Energy recovery from bulk incineration is well established. The constraints to further recovery are associated with net costs; these in turn are dependent on being able to sell the surplus energy at a reasonable price. As just noted when electricity is supplied to the grid there is a tendency for this to be at low prices.

Potential action for the EEC or Governments of Member States to increase the volume of municipal wastes going to bulk incineration would be to:

o       support research and development into extending
        distribution networks and reducing heat losses in
        district heating systems;

o       encourage the use of electricity generated by
        incineration, possibly through subsidies, by electricity
        utilities and electricity companies.

**Small-scale incineration**: The major constraint here is that a regular supply of waste material is not guaranteed. This means that in cases where regular energy requirements are needed, incineration can only act as a supplementary fuel source. This restricts its application and increases unit costs.

Possible EEC/Government action might comprise:

o       where this technology is feasible consider investment
        subsidies to encourage industrial units to utilise their
        combustible wastes for energy recovery;

o       encourage thermal conversion of wood wastes by saw-
        milling and wood converters over and above present
        utilisation aimed at energy self-sufficiency for the
        industries;

o       the problems encountered in the sale of surplus energy
        from wood waste firms should be studied and recommenda
        tions made;

o       provide incentives for energy self-sufficiency;

o       study the problems of sale of excess energy from small-
        scale generators;

o       continued research and development of techniques aimed
        at providing reliable and efficient means of waste to
        energy conversion;

**Refuse-derived fuel:** RDF combustion trials are on-going and supported by the Commission. These should be continued.

Potential action is therefore limited to the following for the Commission and Member State Governments:

o        encourage the establishment of standard specification parameters for RDF pelletised fuel;

o        promote exchange of information on the production, marketing and combustion of RDF;

o        consider providing price subsidies to users to encourage trial by potential users;

o        provide grants for boiler conversions.

**Pyrolysis:** If the technology becomes established a constraint for utilising post-consumer wastes will be providing sufficient quantities of waste for economic operation.

Action by the Commission or Member State Governments might comprise the following:

o        monitor the development of existing 'new generation' pyrolysis plants;

o        provide assistance with research and development as necessary;

o        fund demonstration plants if technical competence is established;

o        investigate methods of waste collection to enable sufficient and regular quantities of wastes to be provided to plants, in particular waste rubber.

**Appendix A**

**WASTE ARISINGS AND MATERIAL RECOVERY RATES
FOR EEC MEMBER STATES**

**COMMODITY: Aluminium**

| Country | Consumption (Mt) | Dissipative Losses Mt | Dissipative Losses % | Physically Available (Mt)(2) | Recovered (+ imports - exports) Mt | Recovered % | Further Recovery Possible (Mt) |
|---|---|---|---|---|---|---|---|
| Belgium | 0.190E (1) | 0.133 | 70 | 0.057 | $0.001 - 0.039 + 0.039 = 0.001$ | 2 | 0.056 |
| Denmark | 0.068 | 0.048 | 70 | 0.020 | 0.014 | 70 | 0.006 |
| France | 0.733 | 0.513 | 70 | 0.220 | $0.161 - 0.012 = 0.149$ | 68 | 0.071 |
| Greece | 0.855E (1) | 0.599 | 70 | 0.257 | 0.250 | 97 | 0.007 |
| Ireland | - | - | - | - | - | - | - |
| Italy | 0.862 | 0.603 | 70 | 0.259 | $0.301 - 0.094 = 0.207$ | 80 | 0.052 |
| Luxembourg INC Belgium | | | | | | | |
| Netherlands | 0.141 | 0.099 | 70 | 0.042 | 0.056 | 133 | [-0.014] (3) |
| United Kingdom | 0.630 | 0.441 | 70 | 0.189 | $0.172 + 0.031 = 0.203$ | 107 | [-0.014] (3) |
| West Germany | 1.041 | 0.729 | 70 | 0.312 | $0.413 - 0.157 = 0.256$ | 82 | 0.056 |
| (Spain) | 0.411 | 0.288 | 70 | 0.123 | 0.035 | 28 | 0.088 |
| EUR 8 | 5.044 | 3.531 | 70 | 1.479 | 1.171 | 79 | 0.308 |

**Aluminium – Notes on Table**

1.    Estimated on basis of population and mean EUR ten consumption per capita.

2.    Calculated on basis of OECD end-use and published life cycle data and published data on recoverable proportions. Estimates vary from 30-60% recoverable.

3.    [-x] indicates Member State recycling maximum proportion possible as defined by this study; additional recovery may be brought about by improvements to recovery technology or designing products for recovery.

COMMODITY: Plastics

YEAR: 1981

| Country | Consumption (Mt) | Dissipative Losses Mt | Dissipative Losses % | Physically Available (Mt) | Recovered (+ imports - exports) Mt | Recovered (+ imports - exports) % | Further Recovery Possible (Mt) |
|---|---|---|---|---|---|---|---|
| Belgium | 1.50 | 0.50 | 34 | 1.00 | 110.00 | 11 | 0.89 |
| Denmark | 0.28 | 0.09 | 34 | 0.19 | 75.00 | 39 | 0.12 |
| France | 2.61 | 0.88 | 34 | 1.73 | 390.00 | 23 | 1.34 |
| Greece | E1.50 | 0.50 | 34 | E1.00 | E 75.00 | 8 | 0.93 |
| Ireland | 0.13 | 0.09 | 34 | 0.09 | 14.00 | 16 | 0.08 |
| Italy | 2.70 | 0.90 | 34 | 1.80 | 200.00 | 11 | 1.60 |
| Luxembourg INC Belgium | | | | | | | |
| Netherlands | 0.65 | 0.22 | 34 | 0.43 | E 32.00 | 7 | 0.40 |
| United Kingdom | 2.10 | 0.71 | 34 | 1.39 | 200.00 | 14 | 1.19 |
| West Germany | 6.32 | 2.13 | 34 | 4.19 | 547.00 | 13 | 3.64 |
| (Spain) | | | | | | | |
| EUR 10 | 17.79 | 6.00 | 34 | 11.79 | 1.71 Mt - 0.067 imported | 15 | 10.08 Mt |

131

## Plastics - Notes on Table

1.  Consumption derived from Eurostat data.

2.  Dissipitative losses calculated on basis of OECD end-use and published life cycle data and published data on available resources for recycling.

3.  Current levels of recovery provided by APME. BPF and Eurostat.

COMMODITY: Paper & Board

YEAR: 1981

| Country | Consumption (Mt)(1) | Dissipative Losses Mt | % | Physically Available (Mt) | Recovered (+ imports - exports) Mt | | % | Further Recovery Possible (Mt) |
|---|---|---|---|---|---|---|---|---|
| Belgium | 1.383 | 0.207 | 15 | 1.176 | 0.271 + 0.150 = 0.421 | | 36 | 0.756 |
| Denmark | 0.842 | 0.126 | 15 | 0.716 | 0.166 + 0.053 = 0.219 | | 31 | 0.497 |
| France | 6.231 | 0.551 | 9 (2) | 5.680 (2) | 1.910 + 0.038 = 1.948 | | 34 | 3.732 |
| Greece | 0.492 | 0.074 | 15 | 0.418 | 0.110 + N/A = 0.110 | | 26 | 0.308 |
| Ireland | 0.245 | 0.037 | 15 | 0.208 | 0.025 + 0.018 = 0.043 | | 21 | 0.165 |
| Italy | 5.051 | 0.758 | 15 | 4.293 | 2.260 - 0.650 = 1.61 | | 38 | 2.683 |
| Luxembourg INC Belgium | | | | | | | | |
| Netherlands | 2.192 | 0.322 | 15 (2) | 1.870 (3) | 0.935 + 0.093 = 1.028 | | 55 | 0.842 |
| United Kingdom | 6.895 | 1.395 | 20 (4) | 5.500 (4) | 2.190 + 0.176 = 2.366 | | 43 | 3.134 |
| West Germany | 9.652 | 1.448 | 15 | 8.204 | 3.250 + 0.115 = 3.365 | | 41 | 4.839 (5) |
| (Spain) | 2.894 | 0.434 | 15 | 2.460 | 1.203 - 0.229 = 0.974 | | 40 | 1.486 |
| EUR 8 | 29.596 | 4.918 | 15 | 28.065 | c.12.000 | | 37 | 16.956 |

## Paper and Board - Notes on Table

(1)     CEPAC data, 1981 Paper and Board Notes on table opposite.

(2)     INSA, Lyon.

(3)     Dutch Govt. figure.

(4)     UK Govt. figure.

(5)     Assoc. of West German Paper Mills, Materials Rec. Weekly, December 18/25 1982.

COMMODITY: Glass

| Country | Consumption (Mt) | Dissipative Losses Mt | Dissipative Losses % | Physically Available (Mt)(3) | Recovered (+ imports - exports) Mt | Recovered (+ imports - exports) % | Further Recovery Possible (Mt) |
|---|---|---|---|---|---|---|---|
| Belgium | 0.50 | 0.22 | 43 | 0.29 | 0.1 | 34 | 0.28 |
| Denmark | 0.42 | 0.18 | 43 | 0.24 | 0.02 | 8 | 0.22 |
| France | 3.49 | 1.50 | 43 | 2.00 | 0.42 | 21 | 1.58 |
| Greece | no data | | | | (virtually none) | | |
| Ireland | 0.13 | 0.06 | 43 | 0.07 | 0.007 | 10 | 0.06 |
| Italy | 2.91 | 1.25 | 43 | 1.66 | 0.35 | 21 | 1.31 |
| Luxembourg INC Belgium | | | | | | | |
| Netherlands | 0.66 | 0.28 | 43 | 0.38 | 0.17 | 45 | 0.21 |
| United Kingdom | 2.82 | 1.21 | 43 | 1.61 | 0.085 | 5 | 1.53 |
| West Germany | 4.65 | 2.00 | 43 | 2.66 | 0.67 | 25 | 1.99 |
| (Spain) | | | | | | | |
| EUR 8 | 15.57 (1) | 6.66 | 43 | 8.91 | 1.82-0.015 (2) | 20 | 7.10 |

**Glass – Notes on Table**

1.  Made up from 9.38 Mt container glass, 3.5 Mt flat, 0.62 Mt glass fibre, domestic 0.73 Mt and all other glass 1.34 Mt.

2.  0.015 tonnes of cullet incorporated

3.  Assumes 95% recovery from container glass with no recovery from other glass consumed.

COMMODITY: Rubber

YEAR: 1981

| Country | Consumption (Mt)(1) | Dissipative Losses Mt(3) | % | Physically Available (Mt) | Recovered (+ imports - exports)(2) Mt | % | Further Recovery Possible (Mt) |
|---|---|---|---|---|---|---|---|
| Belgium | 0.055 | 0.010 | 22 | 0.045 | 0.009 | 20 | 0.036 |
| Denmark | 0.028 | 0.005 | 22 | 0.023 | 0.008 | 35 | 0.015 |
| France | 0.471 | 0.083 | 22 | 0.388 | 0.140 | 36 | 0.248 |
| Greece | 0.053 | 0.009 | 22 | 0.044 | 0.009 | 20 | 0.035 |
| Ireland | 0.019 | 0.003 | 22 | 0.016 | 0.003 | 19 | 0.013 |
| Italy | 0.388 | 0.069 | 22 | 0.319 | 0.060 | 19 | 0.259 |
| Luxembourg INC Belgium | | | | | | | |
| Netherlands | 0.086 | 0.015 | 22 | 0.071 | 0.009 | 13 | 0.062 |
| United Kingdom | 0.340 | 0.060 | 22 | 0.280 | 0.109 | 39 | 0.171 |
| West Germany | 0.565 | 0.100 | 22 | 0.465 | 0.153 | 33 | 0.312 |
| (Spain) | - | - | - | - | - | - | - |
| EUR 10 | 2.005 | 0.36 | 22 | 1.65 | 0.500 | 28 | 1.15 |

## Rubber – Notes on Table

1.  Consumption pattern derived from International Rubber Study Group and Eurostat data. Includes natural and synthetic rubbers, synthetic rubber 68% of total 1981 consumption in EUR 10.

2.  Recovery based on rubber content of reclaiming crumb and retreaded tyres.

3.  Dissipitative losses based on published data for tyres and other products by EEC, RAPRA (UK) and other published data.

COMMODITY: Wood

YEAR: 1981

| Country | Consumption (Mm³)(1) | Dissipative Losses Mm³ (1) | % | Physically Available (Mm³)(1) | Recovered (+ imports − exports)(1,2,3) | | Mm³ | | % | Further Recovery Possible (Mm³) | % |
|---|---|---|---|---|---|---|---|---|---|---|---|
| Belgium | 3.896 | 1.979 | 50 | 1.919 | R | 0.487 | = | 0.974 | 51 | 0.945 | 51 |
| | | | | | E | 0.487 | | | | | |
| Denmark | 1.808 | 0.918 | 50 | 0.890 | R | 0.226 | = | 0.452 | 51 | 0.438 | 51 |
| | | | | | E | 0.226 | | | | | |
| France | 44.451 | 9.723 | 22 | 34.728 | R | 13.891 | = | 27.782 | 80 | 6.946 | 80 |
| | | | | | E | 13.891 | | | | | |
| Greece | 14.320 | 7.275 | 50 | 7.053 | R | 1.790 | = | 3.580 | 51 | 3.473 | 51 |
| | | | | | E | 1.790 | | | | | |
| Ireland | 0.360 | 0.183 | 50 | 0.177 | R | 0.045 | = | 0.090 | 50 | 0.087 | 50 |
| | | | | | E | 0.045 | | | | | |
| Italy | 22.640 | 11.501 | 50 | 11.150 | R | 2.833 | = | 5.660 | 51 | 5.490 | 51 |
| | | | | | E | 2.833 | | | | | |
| Luxembourg INC Belgium | | | 50 | | R | | | | | | |
| | | | | | E | | | | | | |
| Netherlands | 0.760 | 0.386 | 50 | 0.374 | R | 0.095 | = | 0.190 | 51 | 0.184 | 51 |
| | | | | | E | 0.095 | | | | | |
| United Kingdom | 6.510 | [3.307] | 50 | 3.206 | R | 0.305 | = | 0.976 | 30 | 2.230 | 30 |
| | | | | | E | 0.671 | | | | | |
| West Germany | 34.000 | 0 (4) | 0 | 16.745 | R | 9.800 | = | 16.0 | 96 | 0.745 | 96 |
| | | | | | E | 6.200 | | | | | |
| (Spain) | 10.800 | 5.486 | 50 | 5.319 | R | 1.350 | = | 2.7 | 51 | 2.619 | 51 |
| | | | | | E | 1.350 | | | | | |
| EUR 8 | 142.500 | 42.115 | 45 | 80.240 | TOTAL 69.500 | | RESOURCE 36.367 ENERGY 33.133 | | 56 | 21.836 | 56 |

## Wood - Notes on Table

1.      Data from UN/FAO, Geneva.  Medium term survey of trends in the markets for pulpwood, wood for energy and miscellaneous roundwood. Supp. 15 to Vol. 34 'Timber Bulletin for Europe'.

      N.B. Total consumption

| | | |
|---|---:|---|
| = | 44.9 | sawnwood |
| = | 72.5 | industrial |
| = | 22.1 | pulpwood |
| | 3.0 | pit props etc. |
| | 142.50 | |

of which 92.4% was supplied by EEC countries.

2.      Recovered divided between wood wastes as a resource and fuel. Where correspondents to survey did not provide replies UN/FAO estimates of 50% recovery for energy and fuel have been used. Trade as a resource is reduced by low bulk density.

3.      Total includes 10.00 $Mm^3$ of bark recycled to products or energy.

4.      FRG UBA figures state that no wood resources are unrecoverable but process losses may amount to between 0 and 10%.  (Seminar on Energy Conservation and Self-Sufficiency in the Sawmill Industries.  UN/ECE Bonn. 13-17/September 1982.)

**COMMODITY: Ferrous Metals**

| Country | Consumption (Mt) | Dissipative Losses Mt | % | Physically Available (Mt)(2) | Recovered (+ imports − exports) Mt | % | Further Recovery Possible (Mt) |
|---|---|---|---|---|---|---|---|
| Belgium | 4.970 | 1.590 | 32 | 2.883 | − 0.350<br>3.970 + 0.039 = 3.659 (4) | 127 | [−0.776] |
| Denmark | 1.510 | 0.453 | 32 | 0.876 | 0.107 + 0.015 = 0.122 | 14 | 0.754 |
| France | 21.150 | 6.768 | 32 | 12.267 | 8.000 + 2.860 =10.860 | 89 | 1.407 |
| Greece | E1.50(1) | 0.480 | 32 | 0.870 | 0.250 to 0.300 | 32 | 0.570 |
| Ireland | 0.453 | 0.145 | 32 | 0.263 | 0.037 | 14 | 0.226 |
| Italy | 26.970 | 9.590 | 32 | 15.643 | 16.010 − 5.570 =10.440 | 67 | 5.203 |
| Luxembourg INC Belgium | | | | | | | |
| Netherlands | 3.651 | 1.168 | 32 | 2.118 | 1.580 + 1.010 = 2.590 | 122 | [−0.472] |
| United Kingdom | 18.030 | 5.770 | 32 | 10.457 | 8.810 + 3.060 =11.885 | 114 | [−1.428] |
| West Germany | 41.446 | 13.263 | 32 | 24.039 | 17.146 + 2.120 =19.356 | 81 | 4.683 |
| (Spain) | E7.978(1) | 2.553 | 32 | 4.627 | − | − | − |
| | | | | | TOTAL DOMESTIC (2)<br>= 36.149<br>INDUSTRIAL<br>= 22.800 | | |
| EUR 8 | 126.051 | 40.336 | 32 | 73.110 | 58.949 | 81 | 13.250 |

## Ferrous Metal – Notes on Table

1.  Estimated on basis of population and mean consumption per capita
    due to Manisali, E and Birdal, I Demir Celik Sanayinde Gelismeler
    Ve Scrunlar, Aralik 1981.  (**Turkish Steel Ind. Paper**).

2.  Calculated on basis of OECD, Eurostat and published data, at 68%
    of consumption.

3.  Recovery divided between industrial and domestic arisings as
    defined by Eurostat;  'domestic' refers to sources other than
    steel/iron producing industries but may include steel/iron using
    industries;  as defined 'domestic' arisings are therefore post-
    consumer in nature.   Industrial arisings are made up almost
    entirely of circulating scrap, this in turn representing about 18%
    of the EUR 10 crude steel production.
    5% of industrial scrap is capital and arises from the demolition
    of old steelwork plant and equipment.

4.  1980 data.

**COMMODITY: Copper**   YEAR: 1981

| Country | Consumption (Kt)(1) | Dissipative Losses Kt | Dissipative Losses % | Physically Available (Kt)(2) | Recovered (+ imports − exports)(3) Kt | % | Further Recovery Possible (Kt) |
|---|---|---|---|---|---|---|---|
| Belgium | 260.00 | 110.60 | 43 | 39.00 IND. / 149.00 | 58.00 DOM. − 18.00 = 79.00 | 53 | 70.00 |
| Denmark | 1.20 | 0.51 | 43 | 0.69 | + 13.10 = 13.10 | >100 | [−12.41] |
| France | 429.60 | 182.75 | 43 | 116.00 IND. / 246.85 | 33.00 DOM. + 80.50 = 229.50 | 93 | 64.1 |
| Greece | − | − | − | − | 10.00 IND. = 10.00 | − | − |
| Ireland | 0.20 | 0.09 | 43 | 0.12 | − | 0 | 0.12 |
| Italy | 366.0 | 155.70 | 43 | 210.30 | 199.00 IND. − 59.30 / 24.00 DOM. + 6.00 = 169.70 | 81 | 40.60 |
| Luxembourg INC Belgium | | | | | | | |
| Netherlands | 28.00 | 11.91 | 43 | 16.09 | − 18.56 / + 25.10 = 6.50 | 40 | 4.18 |
| United Kingdom | 331.00 | 140.81 | 43 | 190.19 | 128.00 IND. / 76.33 DOM. + 13.20 = 217.53 | 1.14 | [−27.34] |
| West Germany | 744.20 | 316.50 | 43 | 427.50 | 235.00 IND. / 180.00 DOM. − 54.00 = 361.00 | 84 | 66.50 |
| (Spain) | − | − | − | − | 20.00 IND. / 65.00 DOM. − 50.00 = 35.00 | − | − |
| EUR 8 | 2160.20 | 918.95 | 43 | 1240.74 | Total 1121.33 / IND. 648.13 / DOM. 473.20 | 90 | 119.41 / 205.75 |

**Copper - Notes on Table**

1.      World Bureau of Metal Statistics.

2.      OECD Non-Ferrous metal and published data on potential recovery, %
        recovery ranges from 50 to 60% and is highest of all non-ferrous
        metals.

3.      Recovery divided between industrial and domestic arisings as
        defined by Eurostat.   See ferrous metal notes (3).

COMMODITY: Lead

| Country | Consumption (Kt)(1) | Dissipative Losses Kt | % | Physically Available (Kt)(2) | Recovered (+ imports - exports) Kt | % | Further Recovery Possible (Kt) |
|---|---|---|---|---|---|---|---|
| Belgium | 55.00 | 25.25 | 46 | 29.70 | 28.00 + 6.00 = 34.00 | 114 | [-4.30] |
| Denmark | c. 20.00 | 9.18 | 46 | 10.80 | 26.50 - 2.80 = 23.70 | 219 | [-12.90] |
| France | 211.00 | 96.85 | 46 | 113.94 | 99.40 | 87 | 14.54 |
| Greece | c. 15.00 | 6.89 | 46 | 8.10 | 0.70 | 9 | 7.40 |
| Ireland | 10.00 | 4.59 | 46 | 5.40 | 13.00 - 2.20 = 10.80 | 50 | [-5.40] |
| Italy | 265.00 | 121.64 | 46 | 143.10 | 97.40 - 15.27 = 82.13 | 57 | 60.97 |
| Luxembourg INC Belgium | | | | | | | |
| Netherlands | 42.00 | 19.28 | 46 | 22.68 | 19.70 + 22.20 = 41.90 (3) | 185 | [-19.22] |
| United Kingdom | 273.99 | 125.76 | 46 | 147.96 | 206.22 + 17.43 = 223.65 | 151 | [-75.69] |
| West Germany | 331.00 | 151.93 | 46 | 178.74 | 158.00 - 18.00 = 140.00 | 78 | 38.74 |
| (Spain) | 102.00 | 46.82 | 46 | 55.08 | 34.10 | 62 | 20.98 |
| EUR 8 | 1222.99 | 561.35 | 46 | 660.42 | 656.28 | 99 | 4.14 |

## Lead - Notes on Table

1.      World Bureau of Metal Statistics and Metal Bulletin.

2.      OECD Non-Ferrous metal and published data on potential recovery;
        % recoverable ranges from 50 to 60%.

3.      Includes 8.22 Kt remelted, remainder secondary refined.

COMMODITY: Zinc

| Country | Consumption (Kt)(1) | Dissipative Losses Kt | Dissipative Losses % | Physically Available (Kt)(1,2) | Recovered (+ imports - exports)(3) Kt | Recovered % | Further Recovery Possible (Kt) |
|---|---|---|---|---|---|---|---|
| Belgium | 139.00 | 90.35 | 65 | 48.65 | + 11.50 − 6.76 = 11.50 | 24 | 37.15 |
| Denmark | 12.00 | 7.80 | 65 | 4.20 | − | 0 | 4.20 |
| France | 272.00 | 176.80 | 65 | 95.20 | + 7.86 − 8.88 = 7.86 | 8 | 87.34 |
| Greece | − | − | 65 | − | − | − | − |
| Ireland | − | − | 65 | − | − | − | − |
| Italy | 215.00 | 139.75 | 65 | 75.25 | − 6.52 = − | − | [75.25] |
| Luxembourg | | | | | | | |
| Netherlands | 49.00 | 31.85 | 65 | 17.15 | + 14.84 − 6.40 | 87 | 2.31 |
| United Kingdom | 190.00 | 123.50 | 65 | 66.50 | 49.40 + 7.52 = 56.92 | 86 | 9.58 |
| West Germany | 443.00 | 287.95 | 65 | 155.05 | 35.09 + 7.40 − 23.97 = 18.52 | 12 | 136.50 |
| (Spain) | 105.00 | 68.25 | 65 | 36.75 | − | − | 36.75 |
| EUR 8 | 1320.00 | 858.00 | 65 | 462.00 | 109.64 | 24 | 352.36 |

## Zinc – Notes on Table

1.        OECD Non-ferrous metal statistics.

2.        FRG and UK government data in conjuncton with published data on potential recovery rates.

3.        World Bureau of Metal Statistics. World Metal Bulletin Handbook, 1982, Vol. II. (Statistics and Memoranda), Metal Bulletin Publishers, London UK.)

COMMODITY: Tin

| Country | Consumption (Kt)(1,2,3) | Dissipative Losses | | Physically Available (Kt) (1,2,3,4) | Recovered (+ imports − exports)(1,2,3,4) | | Further Recovery Possible (Kt) |
|---|---|---|---|---|---|---|---|
| | | Kt | % | | Kt | % | |
| Belgium | 2.60 | 1.90 | 73 | 0.70 | 0.24 | 34 | 0.46 |
| Denmark | 0.15 | 0.11 | 73 | 0.04 | 0.12 | 300 | [-0.08] |
| France | 10.12 | 7.39 | 73 | 2.73 | - | 0 | 2.73 |
| Greece | 0.05 | 0.37 | 73 | 0.14 | - | 0 | 0.14 |
| Ireland | - | - | - | - | - | - | - |
| Italy | 5.80 | 4.23 | 73 | 1.57 | - | 0 | 1.57 |
| Luxembourg INC Belgium | | | | | | | |
| Netherlands | 5.20 | 3.80 | 73 | 1.40 | 0.18 | 13 | 1.22 |
| United Kingdom | 13.24 | 9.67 | 73 | 3.57 | 1.80 | 50 | 1.77 |
| West Germany | 15.92 | 11.62 | 73 | 4.30 | 6.69 − 5.47 = 1.22 | 28 | 3.08 |
| (Spain) | 4.94 | 3.61 | 73 | 1.33 | - | - | - |
| EUR 8 | 53.53 | 39.08 | 73 | 14.45 | 3.56 | 25 | 10.89 |

**Tin - Notes on Table**

1.  OECD non-ferrous metal statistics, 1980.

2.  Statistical bulletin of the International Tin Council, 1979.

3.  Bulletin statistique du Conseil International de l'Etain, 1979.

4.  World Bureau of Metal Statistics.  World Metal Bulletin Handbook, 1982, Volume II (Statistics and Memoranda), Metal Bulletin Publishers, London, UK.

COMMODITY:  Waste Oils

YEAR:  1981

| Country | Consumption (Kt)(1) | Dissipative Losses Kt (2,3) | % | Physically Available (Kt)(2) | Recovered (+ imports − exports)(2) | | Kt | % | Further Recovery Possible (Kt) |
|---|---|---|---|---|---|---|---|---|---|
| Belgium | 221.00 | 126.00 | 57 | 95.00 | L | 0 | | | 78.00 |
| | | | | | F | 10 + | 7 = 17 + | 18 | |
| Denmark | 79.00 | - | 10 | 80.00 | L | 4 | | | 8.00 |
| | | | | | F | 68 | = 72 | 90 | |
| France | 893.00 | 473.00 | 47 | 420.00 | L | 245 | | | 120.00 |
| | | | | | F | 18 + | 37 = 300 | 71 | |
| Greece | 100.00 | 67.00 | 67 | 33.00 | L | - | | | - |
| | | | | | F | - | | - | |
| Ireland | 41.00 | 24.00 | 59 | 12.00 | L | 0 | | | 0 |
| | | | | | F | 17 | = 17 | 100 | |
| Italy | 690.00 | 397.50 | 58 | 292.50 | L | 100 | | | 92.50 |
| | | | | | F | 100 + | = 200 + | 68 | |
| Luxembourg | 10.00 | 5.50 | 55 | 4.50 | L | 0 | | | 2.50 |
| | | | | | F | 2 + | = 2 + | 44 | |
| Netherlands | 258.00 | 150.5 | 58 | 107.50 | L | 10 | − 10 | | 3.50 |
| | | | | | F | 85 + + | 19 = 104 + | 97 | |
| United Kingdom | 792.00 | 412.00 | 52 | 380.00 | L | 80 | | | 30.00 |
| | | | | | F | 270 | = 350 | 92 | |
| West Germany | 1204.00 | 694.00 | 58 | 510.00 | L | 268 | −? | | 4.00 |
| | | | | | F | 235 + | 3 = 506 | 99 | |
| (Spain) | 416.00 | 280.05 | 67 | 135.95 | L | - | | | - |
| | | | | | F | - | | - | |
| EUR 8 | 4772.00 | 2833.00 | 59 | 1939.00 | | | 1568.00 | 81 | 350.00 |

**Waste Oils - Notes on Table**

1.      Lubricants Statistiques 1981. Centre Professionel de Lubricants, Paris.

2.      ERL. Implementation of Directive 75/439/EEC on the disposal of waste oils. February 1983.

3.      Waste Management Advisory Council. An economic case study of waste oil. Paper No. 3. HMSO, London (UK) 1976.

4.      CONCAWE 9/73. Disposal of used lubricating oil in N.E. Europe. CONCAWE, Den Haag; December 1973. (Adjusted for 1981 vs 1971 consumption reported by CONCAWE).

153

**Appendix A1**

**REFUSE COMPOSITION IN THE EEC**

APPENDIX A1: EEC MEMBER STATES REFUSE COMPOSITION (1), (2)

| Secondary Material | BELGIUM | | FEDERAL REPUBLIC OF GERMANY | | | | | | DENMARK |
|---|---|---|---|---|---|---|---|---|---|
| Paper/Board | 30 | 43 | 27.5 | 28 | 25 | 20 | 24 | 47.2 | 35 |
| Ferrous metal | 4.5 | 9 | 6.5 | 5 | 4.5 | 3.45 | 5.6 | — | 4 |
| Non-ferrous metal – Aluminium | | | | 5 | | 0.45 | | 0.32–0.64 | |
| Non-ferrous metal – Copper, etc. | | | | | | | | 0.08–0.6 | |
| Ceramics and glass | 10 | 9 | 9 | 16 | 17.4 | 11.6 | 8.0 | 10 | 8 |
| Plastics | 5 | 5 | 4 | 7 (3) | 9.5 | 6.1 | 8.8 | 5 | 4 |
| Textiles | 1.5 | 3 | 3 | | 2.9 | 1.5 | 8.9 (4) | 14 (4) | 2 |
| Rubber | — | — | — | 5 | 1.3 | 2.3 (4) | — | — | — |
| Wood | — | — | — | | | | | | — |
| Reference No. | 1 | 2 | 1 | 3 | 4 | 5 | 5 | 6 | 1 |
| NOTES | | Brussels | | Herne, Deutschland ECOBRITT Production | Westfalia | Household | Total Household and Industrial | | |

APPENDIX A1: EEC MEMBER STATES REFUSE COMPOSITION (1), (2)

| Secondary Material | FRANCE | FRANCE | GREECE | IRELAND | ITALY | ITALY | ITALY | LUXEMBURG | NETHERLANDS |
|---|---|---|---|---|---|---|---|---|---|
| Paper/Board | 11 | 35 | | 33 | 24 | 31 | 30 | 25 | 23 |
| Ferrous metal | 3.3 | 5 ⎱ | | 4 ⎱ | 3.5 ⎱ | 3 | 3 | 3.6 ⎱ | 2.9 ⎱ |
| Non-ferrous metal — Aluminium | 0.3 ⎱ | ⎰ | | ⎰ | ⎰ | 1 ⎱ | 1 ⎱ | ⎰ | ⎰ |
| — Copper, etc. | ⎰ | | | | | ⎰ | ⎰ | | |
| Glass | 7.0 | 8 | | 8 | 5.5 | 6 | 8 | 5.2 | 2.9 |
| Plastics | 2.6 | 5 | | 4 | 6.5 | 2.5 | 10 | 4.6 | 6 |
| Textiles | 2.2 | 4 | | 3 | 3.5 | 3 | 5 (5) | 1.5 | 1.8 |
| Rubber | – | – | | – | – | 2.0 (3) | – | – | – |
| Wood | – | – | | – | – | – | – | – | – |
| Reference No. | 7 | 1 | | 1 | 1 | 8 | 8 | 1 | 1 |
| N O T E S | Nancy RRC – incoming refuse | | | | | National means | Milan | | |

GREECE: NO DATA AVAILABLE

APPENDIX A1: EEC MEMBER STATES REFUSE COMPOSITION (1), (2)

| Secondary Material | NETHERLANDS | NETHERLANDS | UNITED KINGDOM | UNITED KINGDOM | SWEDEN | SWEDEN (9) | USA (a) | Ind. EEC Countries | EEC Mean |
|---|---|---|---|---|---|---|---|---|---|
| Paper/Board | 47 | 23 | 30 | 20-33 | 43 | 50 | 44 | 15-50 | 28.76 |
| Ferrous metal | 0.6 | 3 | 9 | 3-11 | 5 | 5 | 7.6 | 3-10 | 5.13 |
| Non-ferrous metal | | | | | 1 | — | 1 | 0.1-1.0 | 1.0 |
| – Aluminium | | | | | | | | | 0.9 |
| – Copper, etc. | | | | | | | | | 0.1 |
| Glass | 2 | 6 | 9 | 7-15 | 5 | 5 | 6.4 | 4-15 | 8.91 |
| Plastics | 6.8 | — | 3 | 2-5 | 10 (7) | 8 | 4.6 | 2-10 | 4.67 |
| Textiles | 3.8 | — | 3 | 2-6 | 2 | — | 3.6 | 2-7 | 2.75 |
| Rubber | — | — | — | — | 1 (3) | — | 2.5 | — | 1.00 |
| Wood | — | — | — | — | — | — | — | — | 1.50 |
| Reference No. | 9 | 10 | 1 | 11 | 12 | 10 | — | 13 | — |
| NOTES | Apeldorn, TNO Ecofuel production | Wijster Fläkt RRC | | WSL - refuse separation analysis programme | Resource recovery study Scandiaconsult AB | Fläkt RRC, Stockholm | N.E. Massachusetts | | Used in this study |

APPENDIX A1: COMMODITY RECOVERY RATES FROM MUNICIPAL WASTES ACHIEVED IN EUROPE

| Secondary Material | FRANCE | ITALY | NETHERLANDS | SWEDEN | |
|---|---|---|---|---|---|
| Paper/Board | 90% | 18% | 80% | 70% | - |
| Ferrous metal | 90% | 92% | 95% | 100% | 99% |
| Non-ferrous metal | | | | | |
| - Aluminium | } 65% | - | - | - | } 50% |
| - Copper, etc. | | | | | |
| Glass | 85% | - | - | - | 90% |
| Plastics | PVC 75% | 25% | 60% | 82% | 25% |
| Textiles | - | - | - | - | 30% |
| Rubber | - | - | - | - | 80% |
| Other materials | - | 5% residuals + RDF | +compost | +compost | |
| Reference No. | 7 | 8 | 10 | 10 | |
| N O T E S | Nancy, RRC | Milan | Wijster Fläkt process | Stockholm Fläkt process | |

REFERENCE NOTES TO APPENDIX A1 TABLES

(1)    All data are percentages w/w and may be either dry
       or wet weights (not usually given).

(2)    All data are exclusive of putrescible and fine
       ( 6mm) fractions.

(3)    Includes leather.

(4)    Includes leather and wood.

(5)    Includes wood.

(6)    1.3% PVC, 0.3% PE and other plastics.

(7)    0.5% PVC, 9% PE, 0.5% miscellaneous plastics.

(8)    0.25% PVC, 4.00% PE, 0.25% miscellaneous plastics.

(9)    Data on Sweden and USA provided for comparative
       purposes. N.B.: High paper content of Swedish refuse.

## APPENDIX A AND A1: REFERENCE LIST

(1)     Commission of the European Communities, Energy from Municipal
        Wastes, 1980/81. Summary Report, 1983.

(2)     Frerotte, J., Fombreet, J.P.: Methane production from domestic
        refuse. Materials and Energy from Refuse. Proc. 2nd Symposium,
        Antwerp, Belgium, 20-22 October, 1981. Ed. A. Buekens.

(3)     Baker, G.P., Sonnenschein, H.: Storable energy from municipal
        solid waste. Materials and Energy from Refuse. Proc. 2nd
        Symposium, Antwerp, Belgium, 20-22 October, 1981. Ed. A. Buekens.

(4)     Berghoff, R.: On the pyrolysis of domestic waste. Materials and
        Energy from Refuse. Proc. 2nd Symposium, Antwerp, Belgium, 20-22
        October, 1981. Ed. A. Buekens.

(5)     Pautz, D.: Obtaining energy from refuse in the FRG. Recycling
        International: Recovery of Energy and Materials from Residues and
        Waste. Ed. K.J. Thomec-Kozmiensky, Berlin 1982.

(6)     Fisher, P., et al: Vacuum refining of steel recovered from
        municipal solid waste. Recycling International: Recovery of Energy
        and Materials from Residues and Waste. Ed. K.J. Thomec-Kozmiensky,
        Berlin 1982.

(7)     Gilcoux, P., and Gony, J.N.: The Revalord process for sorting and
        materials recovery; design, equipment and economics of Nancy full
        scale demonstration plants. Materials and Energy from Refuse.
        Proc. 2nd Symposium, Antwerp, Belgium, 20-22 October, 1981. Ed. A.
        Buekens.

(8)     Nebbia, G.: La Nuova Ecologia, dic. 82.

(9)     Walpot, J.I. Boesmans, B.: Production of solid fuel from Dutch
        municipal refuse. Materials and Energy from Refuse. Proc. 2nd
        Symposium, Antwerp, Belgium, 20-22 October, 1981. Ed. A. Buekens.

(10)    Cederholm, C.: Recyclable materials from domestic waste and waste
        separation systems. Materials and Energy from Refuse. Proc. 2nd
        Symposium, Antwerp, Belgium, 20-22 October, 1981. Ed. A. Buekens.

(11)    Wallin, S.C. and Clayton, P.: Emmissions from the combustion of
        domestic wastes and RDF. Recycling International: Recovery of
        Energy and Materials from Residues and Waste. Ed. K.J. Thomec-
        Kozmiensky, Berlin 1982.

(12)    Enhörning, B. (s): The costs and value of RDF. Materials and
        Energy from Refuse. Proc. 2nd Symposium, Antwerp, Belgium, 20-22
        October, 1981. Ed. A. Buekens.

(13)    Webb, M., Whalley, L.: Household waste sorting systems. EEC DG 12
        Studies on secondary raw materials, Vol. 1, 1979.

**Appendix B**

**MATERIALS RECOVERY**

B.      **MATERIALS RECOVERY**

The existing situation in respect of the main reclamation activities, and their future potential, is summarised below for the
major materials under review.


B1      **ALUMINIUM**

B1.1    **Existing recovery**

The recovery of waste aluminium in Europe is a well-established
industry.  Around 28% of the aluminium processed in Europe is
recycled metal (see Table B1.1).  Only a small part is
irretrievably 'used up' in dissipative utilisations such as metal
salts, organic and inorganic metal compounds and certain surface
treatments.  The remainder is theoretically 100% retrievable and
although this figure can never be achieved in practice, a
considerable proportion is available.  There is every incentive
to maximise aluminium recovery in Europe where only 12% of
consumption was available from local mining production (1980).

| Table B1.1 | | | |
|---|---|---|---|
| BALANCE OF RECYCLING WITHIN EEC IN 1980 (000 tons) | | | |
| Metal Consumption | Metal Production | Recycling in Smelters Foundries & Semi Finished Products Works | Share of Recycling in Consumption |
| 3.905 | 4,132 | 1,097 | 28% |
| Source:  Metallstatistic, 68th volume, Metallgeseschaft AG. | | | |


B1.2    **Sources of Scrap**

A wide range of aluminium materials may be suitable for reclamation including

o       aluminium scrap thrown up during a processing operation
        such as rolling, extrusion, fabrication and machining or

o       discarded because it has completed its useful life.

The characteristics of the scrap are highly varied.  It may:

o       arise in massive units, such as redundant machinery or
        as a flimsy foil canister seal;

o       be wet, dirty, oily or heavily corroded;

o       be anodised or coated with a heavy paint or lacquer
        layer;

o      be fitted with metallic and non-metallic attachments;

o      range from pure aluminium to that of a complex alloy
       containing up to 10% of copper, magnesium, silicon,
       iron, manganese, nickel, zinc, etc.

B1.3      **Secondary Processing**

Once the material has been segregated from non-aluminium scrap
and waste it can either go directly to the smelter or pass
through the hands of several merchants, who sort out and
segregate batch lots with similar characteristics. The material
may be broken up, shredded or baled before it reaches the
smelters. Established systems exist for the identification of
the various types/grades of scrap.

At the smelters the sorting process continues and the scrap is
converted to a manageable form by flame-cutting, fragmentising or
shredding followed by baling or batching into containers. Within
this operation or after it, the material is prepared for
smelting, when contaminants such as ferrous objects, oil, grease,
water and coatings are removed wherever possible. Each batch is
assessed for its metal value and samples of it analysed to
indicate its composition.

Although open hearth furnaces are used in the reclamation of
clean solid scrap, and some electric induction furnaces are used
for melting clean particulate scrap with an even size distribution, the majority of the scrap is melted in large oil or gas
fired forewell furnaces of 15-50 tonne capacity, or rotary
furnaces of 1-20 tonne capacity.

The secondary smelter products have to be of a very high quality.
Therefore, not every smelter can process just any old or scrap
metal. The use of raw materials depends on the equipment and
processing facilities available. Because of the increasing
occurrence of problematic materials, i.e. materials with
undesirable, partly toxic inclusions, specialisation and division
of labour have developed. This explains the important trade
within the EEC with manufacturing scrap, old metal and metalli-
ferous residue. Its quantities and values are considerably
higher than the foreign trade with non-Community countries.

As more than 60% of the industry's aluminium scrap intake is
'new' scrap that originates from fabricating processes, there is
always a base supply available. This type of scrap is moved
within a virtual closed circuit and the majority of it is
reclaimed currently by the secondary industry for the production
of casting alloys, although, due to the high value of primary
aluminium, an increasing proportion is reverting to the primary
wrought alloy manufacturers. It is considered that little
improvement could be made to its overall recovery rate.

'Old' scrap, accounting for approximately 10% of the aluminium
industry's raw material, is where the potential for any further
recovery lies.

**B1.4**      **Future Recovery Potential**

The two main categories of 'old' scrap which offer the greatest
scope for additional recovery are packaging materials (notably
cans and foils) in domestic refuse and discarded capital goods
(notably vehicles and consumer appliances).

**B1.4.1**     **Aluminium cans**

Although the quantity of aluminium in refuse is comparatively
small the growth in aluminium can consumption provides increasingly significant possibilities for aluminium recovery.  The
consumption of aluminium cans is uneven across Europe, ranging
from around 50% in England and Italy, to virtually 0% in France,.
Where aluminium cans are not yet established, the traditional
steel bodied cans with aluminium ends predominate, providing a
less attractive recovery option because of the multi-material mix
(steel, aluminium, lead/tin solder, etc.).

**B1.4.2**     **Scrap vehicles**

The largest source of aluminium in other post-consumer wastes is
undoubtedly wrecked or abandoned vehicles, but with discarded
consumer appliances also providing a substantial source.  Approximately 26 million vehicles are discarded annually within the
EEC.  Some of these are collected by local authorities and others
are handled by processors and scrap merchants who specialise in
dismantling vehicles.  Many vehicles however are abandoned and do
not reach the reclaimers, representing a considerable loss of
valuable resources.

In order to minimise fuel consumption increasing emphasis is
being placed on light weight construction in automobile design
and construction.  This can be achieved through:

o          reduction in material thickness;

o          optimum material substitution (replacement of steel
           parts by light metal and synthetic materials);

o          light-weight design (new design of parts based upon new
           technologies).

The percentage of aluminium and synthetic materials in automobile
construction has been steadily increasing.  In 1980 a sports car
comprised 20% aluminium by weight compared to an average European
car with 3.5% aluminium [1].  However, research undertaken by the
Automobile Technology Research Association (FAT) into the technically possible substitutions of steel parts with aluminium parts
in a middle range car indicated that the 20% is also achievable

---

[1]  Maximilian J. Wutz. Trends of Automobile Recycling; R.I.
Proceedings, 'Recovery of Energy and Material from Residues and
Waste'.

166

in those types of cars. While it is unlikely that the technical
maximum will ever be achieved in practice, the percentage is
likely to increase substantially above current levels.

**B1.4.3    Reclamation of metals**

Automatic shredders, incorporating hammer mills, are used for the
initial processing. The scrap fed in is minced by rotating
hammers and the material containing iron is separated from the
rest of the scrap by lifting electro-magnets. The remaining non-
magnetic product, which consists of 50-60% of non-metallic
materials and 35-50% of non-ferrous metals, has been traditionally separated by hand.

Float-sink methods have now been developed to extract non-ferrous
metals and are likely to be adopted as this content of automobiles increases. A certain percentage of ferric silicate is
added to a water bath to achieve the float condition. The
separation procedure is accomplished in two stages. Magnesium
and rubber float in the first stage, and aluminium in the second.
The remaining fraction consists of a zinc/brass/copper/stainless
steel mixture. The stainless steel can be separated by hand.
Zinc and zinc alloys are then separated from the remaining heavy
metals thermally.

Aluminium's growing penetration of the automobile market coupled
with emerging technology for processing mixed non-ferrous scrap
point towards the increased importance of junk automobiles as a
future source of reclaimable scrap.

## B2        PLASTICS

### B2.1        Existing Recovery

Plastics are recycled by remelting the plastic and shaping it
when molten, either in moulding dies or by extrusion through
slots or by squeezing it between calender rollers. Most types of
plastics are remeltable (thermoplastic), but there is a substantial minority which are not remeltable (thermoset plastics). The
latter cannot be recycled except for energy or (to some extent)
by grinding then as fillers for other non-remeltable plastic
compositions.

Considerable amounts of prompt and process plastic scrap are
recycled in-plant by resin producers and fabricators. The levels
of scrap recycling are particularly high amongst resin producers
where the waste is usually clean and easily identifiable.
Plastics fabricators also re-work their scrap as far as possible
to produce auxilliary products such as small fittings, furniture
pieces etc. Smaller fabricators usually send scrap to an
independent processor who performs ex-plant the reprocessing
function which the bigger fabricator is able to execute in-plant.
Apart from the secondary plastics that are reclaimed in-plant,
the majority of waste plastics are converted to non-critical by-
products. These are not substitutes for established plastic
products but are offered as substitutes for existing products
made of non-plastic materials.

Despite the considerable degree of recycling activity normally
carried out by fabricators, there is nevertheless a significant
amount of process scrap which is too mixed or contaminated to be
cost-effectively reclaimed. Such wastes are usually tipped or
incinerated.

There is also very limited recycling of mixed plastics arising in
domestic refuse.

### B2.2        Sources of Scrap

Plastic scrap available for recovery can arise in several forms [1]:

o        **Scrap in the form of offcuts, rejects, sprues, etc.,**
         arising in the manufacture of plastic products. The
         bulk of this waste material is recycled by blending
         it with virgin plastic, subject to the careful control of
         the levels of contamination in the plastic regrind and
         the deterioration in the physical properties which may
         be caused by repeated thermal and mechanical processing.

o        **Single grades of contaminated plastic** which can be
         collected from consumers or processors; for example,
         used fertiliser sacks, crates, packaging materials and
         in-house contaminated scrap.

---

[1]  M. Bevis, Secondary Recycling of Plastics, Materials in
Engineering, Vol.3, February 1982.

o     **Single grades of in-house or consumer scrap** contaminated
      with metal attachments. For example, laminated
      aluminium-polystyrene sheet, milk bottles, reject
      mouldings or offcuts containing metal transfers,
      electroplate, etc.

o     **Mixtures of two or more contaminated plastics arising as
      industrial scrap.** For example, sweepings from factory
      floors, scrap cables, carpet trim, laminated film and
      containers.

o     **Light industrial scrap and pre-segregated municipal
      scrap** - PET bottle banks, etc.

o     **Mixed contaminated plastics** as found in municipal
      refuse.

B2.3     <u>Secondary Processing</u>

There are already many existing processes available for recycling
of waste plastics and still many more under development. The
principal recycling routes available to plastics which are to
some degree contaminated or mixed are described below. Only a
small proportion of wastes are processed in this way at present
and articles produced tend to have non-critical applications.

**Direct conversion into saleable artefacts** is widely practised -
several hundred Kauferle Remaker flow moulding machines are in
operation and the Klobbie flow moulding machines developed and
operated by Lankhorst Touwfabrieken BV directly convert mixed
waste into saleable artefacts. A substantial secondary recycling
industry based on these machines now operates.

**Comminution** to fine particle size (<500 microns) in order to
reduce the effects of impurity particles on the mechanical
properties and/or appearance of thin-section artefacts. The FN
Industry homogenising extruder, for example, produces pellets
from mixed waste suitable for conversion into relatively thin-
walled artefacts.

**Wash-Separation** A wide range of wash-sink/float processes are in
use and readily available for secondary recycling operations.
For example, wash reclamation processes are offered by Buckau-
Walther, Societa Progettazoni Industrali, etc. BICC (Metals) UK
have adopted a secondary recycling scheme based on a rising
current separator for the separation of mixed cable insulation
into copper, polyethylene and PVC fractions. Cable insulation
represents an attractive source of scrap for secondary recycling
and naturally occurs in large quantities in a few locations as a
by-product from copper recovery operations.

Some of the most important future developments in plastics
separation technology will probably be based on improvements in
selective froth flotation techniques which have been pioneered in
Japan. There are many patents in existence, although processes
based on these patents have not proved to be a commercial success
to date.

**Melt-Separation**  A recent development in new separation technology relates to this route.  Kauferle (Remaker) are manufacturing
a device which is capable of effecting separation in the melt.
It can, for example, separate high concentrations of aluminium
from polystyrene in laminates, or copper from polyethylene in
telephone wire, without the need for filter screens.  The device
separates on the basis of differences in rheological properties,
and it may be possible to extend its application to mixtures of
plastics.

B2.4    **Future Recovery Potential**

There are a number of difficulties associated with the recovery
of plastics which can be summarised as follows [1]:

o       plastics are not readily sortable by visual inspection,
        feel, or apparent density in the hand as are metals,
        glass and paper;

o       impurities in plastics cannot be 'burnt off' by melting
        at high temperatures as is possible with many metals and
        with glass.  Thus plastics laminated to or stuck to
        other plastics, paper, foils and textiles are difficult
        to separate;

o       plastics have low bulk densities so that concentrations
        of plastic scrap occupy a considerable volume;

o       plastics consist of a very wide range of polymer types
        (e.g. polyethylenes, polypropylenes, polyvinylchlorides,
        polyamides, polyesters, polystyrenes, polycarbonates,
        and many copolymers).  Each polymer type occurs in
        hundreds of different grades of varying properties and
        most formulations contain other additives such as anti-
        oxidants, stabilisers, slip agents and pigments.  The
        processing characteristics and properties of these
        polymers, grades and formulations vary very much more
        widely than do grades of metals or glasses.

The need in plastics processing for a consistent performance from
the molten plastic is why most operations employ a single type of
polymer of known and constant formulation.  For recycling ideal
conditions are **a regular supply of clean dry waste of a single
polymer type of known and consistent formulation.**  BPF have noted
the factors which militate against this:

o       **Irregularity of supply:** the very wide range of grades
        and formulations and the intermittent production of most
        plastic items create substantial problems for the
        regular accumulation of any particular grade.  The
        majority of plastics processing operations are
        continuous in nature and their successful use depends on

[1]  Technical Factors Governing the Recycling of Plastics, The
British Plastics Federation, 1979.

accurately matching the processing conditions to the raw material being handled.

o **Cleanliness and dryness**: unless scrap has been very carefully handled it is readily contaminated and wetted – both having deleterious effects on processing and often causing damage to machinery. Even dust contamination can be a serious problem.

o **Single polymer type and constant formulation**: mixtures of polymer types or even of widely different grades of the same polymer type tend to produce incompatibility giving inconsistent processibility, variable flow characteristics, irregular and unacceptable appearance, poor mechanical properties in the finished article and even total failure to reprocess.

As noted, there is a high degree of recycling of clean 'process' scrap of known composition within the converting factory or by sales to other converting plants. The two largest volumes of remeltable plastics in use are polyethylenes and polyvinyl-chlorides. But when mixed they are particularly incompatible for reprocessing and this is a significant limitation on their recycling. Thus while the recycling of individual polymers in a clean state is carried out extensively, the recycling of contaminated and mixed plastics is faced with severe technical constraints. As most wastes not currently recovered are available in the latter form, recycling prospects appear limited.

B2.5     **Hydrolysis**

Large amounts of waste are generated during the transformation of foams into finished shapes, e.g. cushions, seat squabs, this may be up to 20% of the total raw material input. Only small amounts can be directly recycled as carpet underlay and foam agglomerate and consequently several companies, General Motors, Ford Motor Company and Bayer AG have studied alternative uses. Appreciable quantities of high value foam wastes arise in the production of cars.

Polyurethane foam or other polycondensation or polyaddition polymers can in principle be depolymerised by **hydrolysis**. Work in this area is still largely at the pilot or experimental stage.

**Hydrolysis** yields a waterfree polyol or polyamine which can be reacted with isocyanate. Dissolution in hot polyol or treatment with super-heated steam by continuous reaction or in a fluidised bed have both been experimented with. Liquid containing polyether and amine is recovered from the pulp and can be reprocessed. Yields may approach 85-95%.

The economics of the processes are at present rather poor and so no commercial processes are available.

**B3**      **WASTE PAPER**

**B3.1**    <u>Existing Recovery</u>

In the Community as a whole waste paper has grown in importance
as a raw material to the paper industry in the last ten years;
approximately one third of all paper used is recovered for reuse, while waste paper provides some 45% of the raw material used
in paper production. The recovery rate for waste paper is
broadly similar in all Member States except for the Netherlands -
where it is substantially higher. The extent to which waste
paper is used as a raw material by industry varies considerably
between states, from 25% in Belgium (1981) to 62% in Denmark
(1981).

**B3.2**    <u>Sources of Waste Paper</u>

Most of the waste paper used in the Community is collected from
industry and commerce. Separate collection of paper from
municipal refuse is only widespread in Germany and the
Netherlands, although it is carried out in some areas of most
other Member States. Throughout the Community voluntary groups
are of some significance in waste paper collecting although the
activities of voluntary groups have been limited by the effects
of recession on waste paper prices in the last two years.

Waste paper is extensively traded across borders, and most Member
States both import and export waste paper. Trade patterns can
vary quite substantially from year to year although within the
Community, Italy is generally a net importer and the Netherlands
and Denmark net exporters.

**B3.3**    <u>Future Recovery Potential</u>

There are sharp differences in the extent to which waste paper is
used in different paper and board products and in the scope for
further use:

o           waste papers and board are already extensively used in
            the manufacture of packaging paper and board although
            there remains scope for some further use;

o           waste is increasing in importance as a raw material for
            tissues and sanitary paper, the market for which is also
            growing. This may be one of the key areas where use
            could be substantially increased;

o           waste can be used as a major raw material for newsprint
            although in many cases it is not used or forms a relat-
            ively small proportion of the raw material;

o           high-grade waste, e.g. printers' offcuts, is widely used
            in graphic paper. There may be some scope for using a
            relatively small proporion of lower-grade waste paper
            but this is likely to depend on either changing product
            specifications or substantial shifts in the relative
            prices of virgin pulp and waste, making expensive
            processing (e.g. bleaching) of waste paper feasible.

There are a number of technical constraints which restrict the utilisation of waste paper.

**B3.3.1    Contraries**

One of the principal technical drawbacks to increasing use of waste as a raw material in paper and board production is the presence of 'contraries' - substances which prevent or hinder paper production processes.  The two principal categories of contraries are:

o        inks which are not water-soluble;

o        'stickies', i.e. globules of thermoplastic compounds used in binding and coatings.

The problems presented by these two categories are rather different:

o        de-inking technology is well developed but the process adds substantially to costs.  While significant quantities of waste are de-inked for use in news print manufacture, little de-inking is carried out for other types of product;

o        no technologies have been adopted widely in Europe for dealing with 'stickies' although this is a focus of considerable research effort.  Research is being directed at two aspects of the problem;

.        substitution of water-soluble adhesives for the insoluble adhesives widely used at present;

.        treatment processes to separate stickies from waste paper pulp.

At present several commentators argue that the substitution route is the most likely to be successful. Substitution, however, is unlikely to be adopted rapidly except under extreme economic circumstances because of the costs involved for the adhesives and printing industries.

**B3.3.2    Fibre preservation**

If paper is repeatedly recycled its fibres deteriorate such that the physical properties of the final product degenerate, i.e. there is an increasing bond weakness and the individual fibres weaken.  Considerable research has been devoted to physical and chemical treatments aimed at restoring initial properties or improving them.  The methods showing greatest promise, however, are all high cost and are not likely to prove economically viable.

**B3.3.3    Grading**

Having been separated from other materials and contraries, waste
paper must also be sorted according to grade, because mixing low-
grade with high-grade fibre lowers the final quality, e.g. it is
not possible to repulp newspapers with high grade waste and
obtain a quality product.  The sampling and analysing of a batch
of waste paper can now be done effectively using standard
methods. But within specific grades, qualitative sorting remains
a problem and is virtually a manual operation.

Table B3.3(a) indicates the theoretical limits of recycled fibre
content of various types of paper and board by end-use specification.  These limits do not preclude higher waste paper contents
being achieved within the end-uses stated, but this would alter
the quality of the final product and hence give rise to problems
of consumer acceptability.

| Table B3.3(a) | |
| --- | --- |
| TECHNICAL LIMITS FOR RECYCLED MATERIAL FOR PAPER AND BOARD | |
| | Recycle Limits (% waste paper) |
| **Paperboard** | |
| Unbleached kraft | 10-25% |
| Semichemical pulp | 100% |
| Bleached kraft | 5-15% |
| Combination board | 90-100% |
| **Paper** | |
| Newsprint | 100% |
| Office communications | 10-80% |
| Publishing, printing, converting | 10-80% |

**B3.3.4    Other Uses of Waste Paper**

There is extensive recognition by government and industry that
the entire paper and board arising in municipal waste cannot be
utilised for paper and board manufacture, and that other uses for
waste paper need to be identified. (It is of some importance to
the paper and board industry that alternative uses for low-grade
waste are developed in order that collection circuits do not
entirely lapse when demand from their industry is slack.)

Most of the alternative uses (some are listed in Table B3.3(b))
have been developed on a small scale by entrepreneurs without
financial support from government or the paper industry.

| Table B3.3(b) |
| --- |
| SOME ALTERNATIVE USES OF WASTE PAPER |

Currently undertaken commercially:

. as a packaging material without processing

. in moulded pulp (papiermache) products, egg cartons

. as loft insulation

. as animal bedding

. as a component of fibre fuel

. in chipboard

. in flooring materials

. in decorating materials, e.g. wallpapers

Under investigation

. production of ethyl alcohol, lactic acid etc., following chemical hydrolysis into glucose

. biological conversion to protein

At present, there appears to be little prospect of a large market development in new products or low-grade waste paper.

175

B4        GLASS

B4.1      **Existing Sources and Recovery of Scrap**

Cullet is already well-established in glass manufacturing but a
large proportion of this arises in-house. Breakages, trimmings
and rejects from the production of flat glass provide a constant
source of cullet for the initial melt. Since the specification
for flat glass is usually more critical than that for hollow
ware, flat glass manufacturers are reluctant to use cullet from
sources external to their production line, although if it can be
kept uncontaminated, cullet from flat glass-using industries is
often acceptable.

In a similar way breakages and rejects provide the main source of
cullet in the hollow glass manufacturing industry, although users
of glass packaging also constitute an important source. Cullet
from sources external to the immediate production line, known as
'foreign' cullet is used to a greater extent than in flat glass
production.

The use of outside cullet by glass manufacturers is the main
outlet for reclaimed glass. Only small quantities are used for
by-products. Cullet is used in the manufacture of certain types
of glass fibre for insulation purposes and also in minute glass
beads incorporated as a reflective medium in paints for road
signs. Tiles from cullet, waste ceramic products and vitreous
enamels are also manufactured. The process is dependent upon
cullet being consistent in composition and chemically stable.
The market is established but small and brief comments on this
work are given below.

B4.2      **Future Recovery Potential**

B4.2.1    **Road surfacing**

          (a)       **Glasphalt**

Laboratory and field studies have demonstrated that waste glass
can be used satisfactorily as an aggregate in road making
material. During the last few years over thirty experimental
glasphalt strips have been laid in the USA and Canada. It is
claimed tht glasphalt is easily worked, has good reflective
characteristics, useful heat retention properties, and good skid
resistance. Studies are now being concentrated on the special
properties that glass adds to the surface.

          (b)       **Slurry seal**

This is an asphalt sealing coat with glass particles, laid cold
in emulsion form to a thickness of about 6 mm. It is claimed
that the glass content increases skid resistance and extends the
life of the surface when 50% or more of the aggregate is replaced
with glass.

**(c)      Road repair material**

This has been developed from asphalt, waste sump oil, and ground glass. The purpose is to provide a patching material for roads which utilises as many waste materials as possible.

**B4.2.2    Construction materials**

**(a)      Terrazzo**

This is a floor or wall material in sheet form, 6-10 mm thick comprising glass fragments replacing marble chips in grey or white Portland cement. The surface is ground and polished, exposing the fragments and forming a wear-resistant surface.

**(b)      Thixite**

These are large ceramic materials, fabricated from building rubble, waste glass, and a small amount of clay, with the glass acting as binder. The result is a fired material that has structural and decorative uses, with low water absorption.

**(c)      Foamed glass bricks and tiles**

Foamed glass products can be made from ground waste glass mixed with a foaming agent. Low density products can be made in various shapes with good chemical durability, thermal insulation, and sound absorption. In addition, ground glass waste can be mixed with other solid wastes, such as fly ash and sewage sludge, to form glass tiles.

Foamed glass pellets (light weight aggregates) have also been produced for use as a soil conditioner, where they improve moisture retention and drainage.

**(d)      Pipes**

Ground waste glass mixed with a monomer, e.g. polyester, styrene, or methyl methacrylate is cast in a mould and cured. The product is acid resistant and pipes made of it are stronger than those made of concrete.

**B4.2.3    Other uses of glass**

**(a)      Pozzolan**

Ground glass shows promise as a pozzolan additive in Portland cement concrete to control or eliminate deleterious reactions between cement and certain reactive aggregates. If successfully developed this could mean eventual utilisation of glass fragments in concrete.

**(b)**      **Bulk fillers**

Finely ground flint waste glass can be used as a bulk filler for paints.

**(c)**      **Clay sweetener**

Finely ground glass can be used for upgrading poor clay, improving abrasion resistance and strength of bricks and other products.

While the above developments look promising they are unlikely to absorb more than a small proportion of potentially available cullet and continuing research and development work is required to identify additional markets.

**B5**      **RUBBER**

**B5.1**      <u>Reclaim</u>

**B5.1.1**      **Existing recovery**

Reclaim is produced from vulcanised rubber scrap by breaking down
its vulcanised structure by the action of heat, chemicals and
mechanical work.  Though reclaim has the plasticity and appearance of a new unvulcanised rubber compound it is different in
that the original sulphur network is only partly removed or
modified and the molecular weight of the main chains is reduced.
Moreover it may contain residues of the reclaiming chemicals and
- if tyres have been used - of the fibre reinforcement.

As a result reclaim compounds have lower physical properties than
compounded new rubber and the main reasons for their use are
price and improved processing of rubber compounds.  This price
advantage may increase with rapidly increasing costs of synthetic
rubbers relative to natural rubbers.  Synthetic rubber is now
roughly the same price as natural rubber, and raw about half of
these prices.

The main processing advantages claimed can be summarised as:

o      shorter mixing times;

o      low power consumption;

o      low heat development

o      faster processing on extruder and calenders

o      low swelling and shrinkage of the unvulcanised compound;

o      faster curing of the compounds.

In many rubber products 20-40% of reclaim can be added to the new
rubber content without serious effects on physical properties and
much higher amounts are used in economy products like car mats.
However, traditionally tyre carcass compounds have been the main
outlet for reclaim because of its processing advantages, despite
their relatively low reclaim tolerance which limits the proportion of reclaim to around 10%.

The advent of radial ply tyres has made reclaimed rubber even
less desirable from a technological standpoint.  Nevertheless,
reclaim can be used for a variety of less technically demanding
products such as hoses, fan-belts, etc., but demand for such
applications is limited relative to the potential supply of
reclaimed rubber.

**B5.1.2    Reclaim processes**

The main qualities of rubber reclaim are:

o        whole tyre reclaims - car and truck tyres;

o        butyl reclaims - from car inner tubes;

o        EPT reclaim - EPT rubber profiles.

Many reclaiming processes have been developed.  In all these processes the scrapped rubber has to be shredded and ground into crumb to permit a proper reaction of chemicals and swelling agents with the vulcanised structure, to promote good heat transfer and to remove the fibres by mechanical or chemical action.

Many chemicals are available as reclaiming agents, such as phenol alkyl sulphides and disulphides and various unsaturated compounds and solvents like coal tar and petroleum napthas and tar oil derivatives.

Sodium hydroxide and metal chlorides are used to destroy the fibre fraction.

The market for reclaimed rubber, being based mainly on the processing of discarded tyres, is characterised by a large number of tyre dealers or service stations supplying used tyres either to a few reclaim producers, to large tyre manufacturers for retreading or to numerous small, independent retreaders.  A few private companies purchase used tyres from dealers for 'splitting' purposes, but these do not absorb any significant quantity of used tyres.  Several of the large tyre and rubber goods manufacturers operate subsidiaries for the production of reclaim, particularly in France, Germany and the UK.  Independent reclaim producers are now few in number, with the likelihood of their diminishing even further.  There is no problem of supply of used tyres, with processors and retreaders able to get more than enough to meet their needs.

**B5.1.3    Future prospects**

The reclaiming industry in Europe is faced with a number of problems:

o        A decreasing market in the rubber industry:  tyre carcasses are a declining outlet because the number of plies is decreasing and less reclaim is tolerated in radial tyres.

o        Competition of low priced synthetics from COMECOM countries.

o        Many specifications do not allow incorporation of reclaim in rubber compounds.

o       The large investments needed for more economic installa-
        tions.

o       Transport costs for scrapped tyres: the reclamation
        industry is becoming increasing centralised, hence
        distances travelled to collect used tyres are
        increasing.

o       The financial burden of environmental restrictions.

o       The low value of recovered textile and wire. Copper-
        plated wire is deleterious to rolling and drawing
        operations.

o       Unavailability of suitable peptizers for recent adopted
        rubber compositions.

o       Traditional non-tyre applications of 'reclaim'
        increasingly taken over by plastics.

o       Steelcord tyres are more difficult to process and have
        taken a larger market share.

There is no immediate prospect for the declining use of reclaim
in tyre manufacture to be overcome. At the same time, the use of
reclaim in non-tyre goods will continue to be influenced by the
prices of virgin rubber, cheap fillers (clays and whiting) and
extenders (aromatic and naptheric oils).

Alternative markets need to be found, or created, if the full
production potential of reclaim is to be realised. Research in
the USA has been carried out on the application of reclaim in the
inner liners of tubeless tyres and as an addition in binders,
adhesives etc. The break-down of rubber through high energy
radiation is being studied in France, possibly leading to a new
process for the production of reclaimed rubber. Research
programmes were recommended in the EEC study on rubber waste
recovery (op cit) covering quality control, the influence of
reclaim on existing rubber compounds and on new machinery for
shredding steel braced radial tyres. In the long run such
projects should help to open up a wider market for reclaim.

B5.2    **Retreading**

B5.2.1  **Existing recovery**

Most non-tyre products are discarded at the point where they are
too worn, perished or otherwise damaged to serve their purpose
and clearly the potential for re-use of such items is virtually
nil. The most widely practiced re-use of a rubber product is
that of tyre retreading. In the retread process the worn tread
is replaced by new rubber to recondition the tyre for a second
(or even further) lifetime.

It is the most direct route in rubber recycling because the complicated construction of the tyre is not destroyed and because the amount of rubber lost by road wear is relatively low. Estimates of this amount vary from 8-20% of the original tyre weight. As retreading prolongs the useful life of the tyre it decreases the number of scrapped tyres as well as the consumption of energy and raw materials for the production of new tyres.

In most EEC countries retreads of truck tyres have a share of about 50% in the replacement market with tyres being retreaded two or three times. This is an accepted and reliable operation and is close to its maximum potential. For car tyres the share of retreads is very much lower, as indicated in Table B5.2(a).

The table shows that in several countries of the EEC retreads of truck tyres have a share of around 50% in the replacement market, reflecting the practise of retreading large tyres two or three times and even more. The much lower share of retreading among car tyres reflects:

o        the lower relative price differential between new and retreaded tyres;

o        the greater vulnerability of the casing to damage (requiring more inspection);

o        consumer attitudes which regard retreaded tyres as inferior quality.

| Table B5.2(a) | | | | | | |
| --- | --- | --- | --- | --- | --- | --- |
| RETREADS IN PERCENTS OF REPLACEMENT SALES | | | | | | |
| | France | BRD | Holland | Italy | England | USA |
| Car tyres | | | | | | |
| - new | 95 | 73 | 85 | 73 | 72 | 78 |
| - retreaded | 5 | 27 | 15 | 27 | 28 | 22 |
| Truck tyres | | | | | | |
| - new | 51 | 51 | 51 | 50 | 65 | 62 |
| - retreaded | 49 | 49 | 49 | 50 | 35 | 38 |

## B5.2.2    Retreading processes

There are a great number of retreading processes in operation and they can roughly be divided into two main groups:

o        processes using unvulcanised new rubber, and

o        processes using prevulcanised treads.

**Unvulcanised** rubber is mostly used in the form of 'camel-back' which is an extruded strip of the required cross-section suitable to cover the area to be retreaded. The casing is coated with an adhesive, the camel-back is uniformly applied, the two ends are linked and the whole is carefully pressed onto the casing to remove entrapped air.

Instead of using camel-back it is also possible to apply the new rubber directly from the extruder onto the casing. As the extruder presses the hot rubber onto the casing, a better adhesion is claimed and no adhesive is needed.

The use of the **prevulcanised** treads is of growing importance, especially for large tyres and radial tyres. Because the tread is already cured and only a thin layer of adhesive has to be activated for bonding, the actual retreading operation can be performed at lower temperatures and in shorter times.

It is claimed that overcure of the carcass is minimised in this way. At the same time better mileage is obtained because the freedom in compounding the treadrubber is not limited by flow requirements during retreading.

The major technical problems in retreading are related to the complexity of tyre sizes, tread widths and the variable expansion of the tyre casing during its original use. This forces the retreader to maintain a large inventory of moulds. Tyre design is related to wear resistance, safety, comfort and vehicle construction; tyres are not designed for the purpose of retreading. At present there is a change to radial types with a trend to lower section heights and wider treads. The changes are facing the retreaders with problems as diagonal casings become obsolete and radial casings need adapted equipment; many have not been prepared to adapt their equipment initially although after a period of adjustment this is likely to come about given reasonable market conditions.

B5.2.3    **Future prospects**

Any increase in retreading activity depends on:

o       adaptation of retreading equipment to handle radial tyres;

o       reliable and economic supply of retreadable worn casings;

o       changes in consumer attitudes to retreaded tyres;

o       a sufficiently attractive price differential between new and retreaded car tyres.

The main potential clearly rests with car tyres. In the EEC study on rubber waste recovery (op.cit) it was calculated that a maximum of 22% of EEC tyre consumption might be covered by retreaded tyres. This calculation assumed that worn tyres on wrecked cars in scrap yards and at unknown locations would not be available or suitable for retreading. It was further assumed that only 50% of the remaining worn tyres would be retreadable. On this basis savings of 64,000 tonnes annually would be achieved.

The retreading proportion could also be increased by increasing the minimum tread depth required legally in the Member States. Wear below a certain level of tread can severely damage the vital reinforcements. An increase to 3 mm minimum legal tread requirement could double the percentage currently acceptable for retreading.

## B5.3   Rubber Crumb/Powder

### B5.3.1   Existing situation

Numerous examples exist of solid by-products derived from waste rubber, and particularly from waste tyres. Rubber crumb, which is ground vulcanised rubber, essentially a precursor to reclaim production, is extensively used in the manufacture of carpet underlay, industrial flooring, sports surfaces, adhesives and is occasionally used in road surfacing and other bituminous surface applications. Research is also being conducted into the possibility of using crumb for the manufacture of shoe soles.

Small amounts of powdered rubber scrap have traditionally been recycled into the rubber industry and new markets are growing in sporting and sound-proofing applications. The requirements of these markets are fulfilled with by-products and intermediate products of the retreading and reclaim production processes.

### 5.3.2   Processes to produce rubber crumb/powder

Production of rubber crumb or powder is an energy intensive mechanical process and much work has been conducted in trying to establish suitable technologies with lower energy consumption. One major development has been cryogenic grinding which includes immersion of the tyres in liquid nitrogen and fragmentation immediately thereafter in a conventional hammer mill. At those low temperatures the rubber shatters and produces a finer grade of material for a smaller input of energy.

As with reclaim production all waste rubber products are suitable for production of crumb or powder.

### B5.3.3   Future possibilities

Existing possibilities are summarised in Table B5.3(a).

No quantitative data are available about the potential amount of ground or reclaimed scrap rubber in existing and new rubber products. Much research and development work has still to be done, particularly in the field of extremely fine powders. Even if - for technical or economic reasons - no more than 10% of the rubber now being scrapped could be returned to its origin or to other rubber products, this amount would be a useful contribution to the problem of rubber scrap removal as well as an attractive and direct way of recycling.

---

[1]   Paper presented at the IISRP 23rd Annual Meeting, New Orleans, Lou. USA. April 22nd 1982.

Table B5.3(a)
COMPARISON OF THE CLASSES OF RECYCLING PROCESSES FOR RUBBER

| Class of Processes | Potential Raw Materials Savings | Potential Market | Economics | Advantages | Disadvantage or Barriers to Development |
|---|---|---|---|---|---|
| Retreading | Optimal retreading replaces 64,000 ton of new car tyres (rubber, chemicals, energy, steel wiring etc.) | Large | EUA 23,000,000 annual savings in total or EUA 350 per ton of additionally converted waste. | Real recycling: mileage retreaded car tyre = mileage new tyre. | Image of retreaded car tyre and marketing need improvement. |
| Reclaiming | Optimal reclaiming saves 75,000 ton of waste, replacing raw materials (raw rubber chemicals). | Large | EUA 25,000,000 annual savings in total or EUA 355 per ton of additionally converted waste. | Improves rubber processing - replaces raw rubber in rubber articles. | Image of reclaim and marketing need improvement. |
| Pyrolysis | Energy (solid and liquid fuel) or compounds perhaps | Unknown as market not well defined. | Good if use of products in rubber industry or similar is possible. | Obvious if real recycling of char residue is possible. | Market to be found. |
| Incineration | Energy | Large | Very good if no new investment is needed; medium in other cases. | Immediately possible and saves some fossil fuel. | High investment, low profit and low savings. |
| Road Surfacing | Stone or bitumen (according to chosen process) | Large | Promising for bitumen substitution since generating long life surfacing. | High potential profit by making a good quality surfacing. | Unknown for bitumen substitution. Technical and economical barriers for stone substitution. |
| Sportsground Surfacing | | About 50,000 t/year in EEC | Very good | Safety, comfort, profit. | |
| Artificial Reefs Building | Increase of available sea food | Unknown | Expensive disposal. | Perhaps positive environmental impact. | Environmental impact not perfectly known. |
| Absorption of Pollutants | Positive environmental impact | Several 10,000 t/year in EEC? | Unknown | Efficiency | Technical research still needed before large-scale application. |
| Protein Synthesis | Protein? | | | | Difficulties of developing an efficient process. |

185

More systematic research in this field is needed to find the
optimal combinations of scrap particle sizes, of types and
amounts of new rubber and of other compounding ingredients.

B5.4    **By-products**

The final group of by-products from rubber is the miscellaneous
assortment of products derived from 'tyre-splitting', such as
boat fenders, rubber sandals, crash protectors, fish-reefs etc.
The nature of these products give rise to only a very limited
demand.

B5.5    **Thermal Treatment**

5.5.1   **Existing situation**

Thermal treatment of waste rubbers can comprise of either:

o       pyrolysis, or

o       depolymerisation,

the latter requiring a solvent or medium which can be heated and
acts as a depolymerisation agent.

Both treatments have the potential to treat a larger proportion
of the rubber waste arisings, and a number of potentially viable
pyrolysis units have been constructed in Europe, America and
Japan. Depolymerisation is a newer technology and not at the
same stage of development. Both technologies are capable of
producing materials that can be substituted in rubber production
and the Japanese are conducting research into specific production
of carbon black for rubber tyres by pyrolysis. There is,
however, a degrading of materials on treatment and substitution
for primary materials by thermal processing of secondary
materials is unlikely to be widespread. Pyrolysis oils and gases
do have alternative uses but in many cases the principal use will
be as a fuel substitute.

Appendix D, Section D.3, comprises a review of the latest
technological developments in the field of rubber pyrolysis
worldwide.

B6          **WOOD**

B6.1        <u>Logging Residues</u>

B6.1.1      **Existing recovery**

There is currently considerable interest within the forest
industry in Europe in increasing the total volume of wood removed
at harvest. At present about 22% of the total biomass remains in
the forest following harvesting. Existing recovey is practically
zero.

B6.1.2      **Developments in increasing recovery**

The most significant developments in increasing recovery of wood
from forests are likely to arise from greater deployment of stump
removal machinery. Between 5 and 10% of the tree volume remains
in the ground following conventional harvesting consisting of
stump and roots greater than 5 cm in diameter. Mechanical
removal is capable of removing up to 55% of total stump and root
fraction giving an extra 3-6% yield.

The reason why these residues are left is that they have too low
a value in relation to the cost of harvesting and at current
energy prices it has not been possible to utilise these resources
as a fuel save from some specialised situations. Further
constraints are:

o          fraction of narrow stumps; stumps below 20 cms have to
           be left;

o          hardwood stumps have minimal value as a material;

o          steep slopes or inaccessible terrain;

o          size of area requiring clearance (in Sweden the limit is
           set according to geographical region and distance from
           converters but varies between 5 ha in northern Sweden to
           1 ha in southern Sweden);

o          soil fertility (whole tree utilisation can have a
           significant impact on the nutrient cycle).

Around half of the waste removed could be used for fibre production for the paper industry.

Other logging wastes consist of 40% wood, mainly from branches,
25% bark and peeling wastes, and 35% leaves or needles. Needles
or leaves are a major constraint on the use of these materials
for recovey as opposed to thermal conversion. The percentage of
wood will increase if the cuttings (slash) are left in the forest
so that leaves/needles fall off or increase during handling and
transport for the same reasons. No mechanical techniques exist
for wood segregation from slash and together with constraints of
steep, inaccessible or infertile sites, reduce the volume to 25%
total available. Of this the wood fibre content is around 60%.

Other forest wastes are:

o       thinnings
o       cleanings
o       other fellings.

Recovery of thinnings is generally low and is constrained by
handling practices and debarking technology (drum barking has
been experimented with), together with constraints due to the
site nature and position and the content of contaminants.

Cleanings result from the clearance of residual trees in thinning
operations of which a considerable portion is hardwood.  The same
constraints on recovery of thinnings apply to cleanings.
Assuming that similar technologies apply on debarking and
realisation of wood fibre content then a fraction could be gained
for the fibreboard industry but an equivalent volume will be
produced which could only be used as a fuel.

An alternative approach which has been adopted by some
Scandinavian countries is to utilise machines for reducing low
value timber to chips at the roadside or in the forest.  A major
advantage being the relative cleanliness of the material when
chipped immediately.  The principal outlets for these materials
would be the particle board industry.

**B6.1.3    Future prospects**

In general the additional quantities of material realisable from
whole tree utilisation are relatively expensive and often of
inferior quality to trunk timber.  In some cases it may be so low
as to make extraction and conversion uneconomic.  Continued
development of equipment and systems will reduce costs but the
intractable problems of site nature and accessibility remains.
The main area of interest will probably be stumps which can be
mixed with normal wood at levels of up to 20% on production of
quality sulphate pulp.

Considerable volumes of material will be generated which can only
find use as fuel substitutes.  Unless conversion to solid or
liquid fuels at landing sites can be carried out it is unlikely
that substitution of wood for fuels will prove economic due to
the costs of transporting such bulky materials.  It is considered
that in many regards use for energy recovery may prove the more
practical.

**B6.2       Sawmill Residues**

**B6.2.1     Existing recovery**

The reduction of an unbarked log - geometrically a tapered
cylinder - into square edged parallel sided lumber product
results in the generation of a significant volume of residues;
more than from any other stage of wood production.  The principal
qualities of residues are:

o       green-mill residues
- bark via debarking or in rejected timber;

o       solid-wood residues
- sawdust/woodflour
- pieces of wood as off-cuts broken-down by de-fibrification through mechanical action, e.g. planing or chipping. Generally wastes produced are known as: off-cuts; slabs; edgings;

o       dry mill residues
- sawdust, shavings and woodflour;
- trimmings and rejects;
- contaminated wood residues (glued, coated, painted, nailed or otherwise treated).

Green-mill residues arise from the primary conversion of logs to lumber and are characterised by a high moisture content. Dry-mill residues are generated on further processing of dried lumber and usually possess moisture contents of less than 20% W/W.

A trade-off exists for the sawmill operator:

o       produce high-yield of sawn wood with inclusion of a larger proportion of low-grade and small-size lumber with minimal waste;

o       produce lower-yield, higher value product with greater generation of waste.

Traditionally sawmills operating at greater geographical distances from markets (or within countries with well-developed residue processing industries) operate the latter system. In Europe the current level of sawmill residues varies between 45 and 60% of input timber, but the volume is decreasing due to in-house recovery. Currently about 50% of these utilised residues go for energy recovery and this proportion is likely to increase. The remainder goes to conversion by the wood particle board industries and other miscellaneous uses, e.g. horticultural, agricultural and domestic.

There is also a considerable amount of effort being directed at **reducing** the volume of sawmill residues generated, e.g. low kerf bandsaws or oven knives, improved feed patterns and log-sorting and sizing procedures. Major reductions have also resulted from more efficient debarking. The most significant development has been the development of the chipping headrig method of producing lumber which reduces sawdust generation at the expense of greater but more usable quantities of chip.

Few technological problems exist in utilisation of sawmill residues but use is greatly constrained by collection and transport.

Major uses of sawmill residues, apart from bark, are:

o       pulp and paper making;

o　　　　particle boards:
- woodchip
- fibre
- cement board composites;

o　　　　charcoal;

o　　　　activated charcoal;

o　　　　agricultural, poultry or animal bedding;

o　　　　woodflour, filler in plastics etc.,

o　　　　building blocks and bricks to aid airing or firing.

Wood residues must be clean to obtain reasonable market prices. Value is severely constrained by production/consumption of particle board, and local conditions and transport.

As opposed to materials recovery an equivalent volume of material is recorded for its energy content providing up to 35% of total national requirements for sawmilling activities including drying in France. Due to the constraints in marketing of wood wastes generated by sawmills and the considerable volume of unsuitable materials, e.g. bark, it is likely that heat recovery is likely to absorb the majority of incremental recovery. In view of the significant volume of materials generated, their unit energy value and use of energy by sawmillers there is likely to be an excess of energy generated which could have a significant impact on energy recovery levels.

## B6.2.2　　Developments in increasing recovery

Apart from on-site conversion of wastes to energy further use for other energy or materials is heavily constrained by collection and transport costs. Considerable emphasis has therefore been placed on reducing transport costs by volume reduction of materials, i.e. baling or briquetting.

If wastes could be made available for materials recovery at economic prices then technological development could absorb the extra materials available. These include:

o　　　　pulping of heterogeneous wood particle mixes;

o　　　　separation of bark and chip to produce improved quality chip (can only occur at large scale);

o　　　　conversion of bark into phenolic resins or chemicals;

o　　　　conversion of wood wastes to chemical substitutes, e.g. sugars, alcohols, proteins and other organic raw materials.

Due to both the saturation of current markets, state of present technology for chemical conversion of wood wastes and transport factors, it is unlikely that materials recovery will be viable on a wide scale. Conversely the equipment to store, convey and burn wood wastes or thermally convert the materials is available at a range of scales. Additionally environmental standards can be met by existing equipment. Bark still remains a challenge in terms of efficient conversion but briquetting and specialist kiln developments will probably overcome difficulties of moisture content, ash and calorific value variation.

**B6.2.3    Future prospects**

The relative quantities of sawmill residues generated are likely to decrease in future. By far the most likely scenario would be increased energy conversion of the sawmill wood wastes generated. Major constraints would be external markets for fuel substitutes produced or surplus energy generated.

**B6.3    Pulpwood Residues**

**B6.3.1    Existing and potential uses**

Bark is the principal waste arising from pulpwood conversion and is a major problem. Existing levels of recovery are low and increasing recovery will be constrained by:

o       identifying further markets for material recovery;

o       difficulties in conversion to energy;

o       transport costs;

o       present status of technology for conversion of bark to chemical substitutes.

Prospects for chemical conversion include adhesives and resins respectively for wood particle boards exist but are not commercial. Markets for direct use exist such as potting mediums, soil improver, horticultural mulch and peat substitute but these are limited by transport costs.

The most likely use for bark remains energy conversion either directly or by sale as a briquetted fuel substitute.

**B6.4    Timber Conversion**

**B6.4.1    Existing and potential uses**

Existing and potential uses for timber conversion wastes are the same as for sawmill residues. Timber conversion wastes may have higher value due to increased homogeneity but may be produced in smaller and more fragmented units making recovery more difficult. Use of timber conversion wastes for energy generation is likely to prove the most viable option.

**B6.5**    <u>Post-Consumer Timber</u>

**B6.5.1**   **Existing and potential uses**

Post-consumer timber is defined as wood products which are no longer used for their original purpose and life-use has expired. This category includes a number of solid wood products, e.g. pallets, boxes, crates, railway sleepers and construction timber, e.g. scaffolding planks and shuttering. Apart from demolition wastes there is no major aggregation of post-consumer wood wastes. Estimates exist that about 10-15% of construction timber is wasted following its useful life.

Apart from direct reuse the use of post-consumer wastes will probably be restricted to incineration if suitable quantities arise en masse.

Appendix C

**MATERIAL RECOVERY SYSTEMS**

C.      **MATERIAL RECOVERY SYSTEMS**

A common feature of the materials being reviewed here is that, apart from rubber and wood (and small quantities of plastics), the potential for further recovery has been identified as being from domestic and trade wastes. The crucial question thus becomes, how can recovery from these sources best be achieved? There are two main options, although each has several variants:

o       mechanical separation of materials from mixed refuse at central locations;

o       material sorting at source (in homes, restaurants, offices etc.).

In this section we review experience to date in the EEC of these alternatives, consider their respective advantages and disadvantages and assess likely future developments.

C1      **Mechanical Sorting and Separation**

Attention is concentrated here on mechanical systems which can recover aluminium, plastics, glass and paper for recycling as materials. The recovery of ferrous metals, RDF fractions and organic fractions for compost are not considered, except where this occurs in conjunction with one of the above-mentioned materials.

C1.1    **Technical approach**

Numerous recycling methods have been developed in recent years to reclaim components of mixed wastes by means of automatic sorting processes, followed by additional treatments to purify and convert the separated portions to produce different products suitable for commercialisation.

It is not within the scope of this study to undertake a detailed technical evaluation of alternative mechanical recovery systems. Rather, it is to assess the efficacy of such systems, their energy requirements and likely future developments.

The dominant characteristic of household waste is its heterogeneous nature; its components possess a wide variety of sizes and shapes and physical and chemical properties. Separations are usually achieved by the identification and use of a particular component property and many different principles have been tested. These include size, shape, mass, colour, density, friction, elasticity, magnetic properties, electrical conductivity etc. In practice, although similar properties are exploited for separation, designs of equipment are variable as is the order in which components are removed from the bulk waste stream.

A technical description of household waste sorting systems is available in the recent CREST report [1]. Based on this we summarise below the main approaches being adopted for the recovery of plastics, aluminium, paper and glass.

C1.2    Recycling plants in Europe

In spite of the great interest in the recovery of materials from mixed wastes and the experimental work undertaken, the plants currently operating on a commercial scale are very few in number. In addition some plants which started out with the objective of material recovery have since switched to producing RDF, compost or have shut down. The plants concerned with material recovery at present are reviewed below.

Rome, Italy

There are two recycling plants in Rome utilising the Sorain-Cecchini method. An initial selection is carried out, based on screening, using different meshes, and air sorting. The separate portions are stabilised by additional treatments which result in a pulp of paper waste, detinned scrap iron and for plastic. Compost is also made and residual combustibles are incinerated to provide steam for the sterilisation and other treatment processes. The process has been operational since 1964. The composition of the incoming waste and the materials recovered are shown in Table C1.2(a),

| Table C1.2(a) | | | |
|---------------|---|---|---|
| % MATERIAL RECOVERY AT ROME PLANT | | | |
| Material | Refuse<br>% Composition | Material | Material<br>% Recovery |
| Ferrous Metal | 3 | Ferrous Metal | 3 |
| Plastic | 4 | Plastic | 2 |
| Paper | 18 | Paper | 13 |
| Glass | 4 | Compost | 24 |
| Organic | 50 | Organic to<br>animal feed | 16 |
| Miscellaneous | 21 | Incineration and<br>steam production | 42 |

[1] Studies on Secondary Raw Materials: Vol 1 Houeshold Waste Sorting Systems, January 1979 - EEC.

The percentage of materials recovered are average values; the
flexibility of the process permits by-passing where necessary any
product recovery where a market does not temporarily exist.

## Paper

The baled paper resulting from sorting is pulped, cleansed and
dewatered to remove contaminants and to provide a commercially
acceptable product. Further processing is undertaken to
eliminate contraries such that paper-pulp for both cardboard and
newsprint is produced.

## Plastic

Plastic film, made up of low density polyethylene, is the main
type of plastic in Rome's waste and is recovered. Other types of
plastic are not recovered. Originally the plastic film was
salvaged and baled and sold to the injection and compression
moulding plastic industry. Sorain-Cecchini have now installed
plant to convert plastic film to granules which have sufficient
purity to allow blending up to 50% with virgin resin for the
production of plastic film.

## Glass

The recovery of glass is still at an experimental stage. It is
intended to colour-sort glass and modifications are being made
which it is hoped will result in technical and economic
viability.

## Non-Ferrous Metals

The percentage of non-ferrous metals in refuse collected in Rome
is almost negligible and is not sufficient to justify recovery.

## Neuss, Federal Republic of Germany

The first large-scale material recovery plant operating in the
Federal Republic of Germany is located at Neuss in Westphalia.
It has been developed by M. Trienekens Company and constructed
with the assistance of a Federal Government grant (37%).

It is situated at a landfill site and the refuse can be directed
either to the plant or to the landfill. The throughput of the
plant is 60 t/hr of household wastes and 6.5 t/hr of industrial
refuse.

The commercial refuse from industry handled in the plant consists
mostly of packing material. Various paper qualities are hand-
picked, shredded and baled.

The household refuse is first screened to separate organic
material, glass, ashes etc. Plastic foils are mechanically
separated and the oversize materials, after shredding, magnetic
extraction and further screening are air classified into light

and heavy fractions. The light fraction consisting of more than 90% paper is baled. Glass and non-ferrous metals are also reclaimed. The annual recovery potential of the plant has been given as follows:

| | | |
|---|---|---|
| Paper | 18,000 | tonnes |
| Glass | 2,500 | " |
| Ferrous | 5,000 | " |
| Non-ferrous | 200 | " |
| Plastics - from trade | 600 | " |
| - from household | 4,000 | " |

An organic fraction is also recovered and worked into compost which is then used as landfill cover material.

About 50% of the mass input can be recovered which is equivalent to a volume saving of 70%.

The plant started up about two years ago and is reported as operating successfully, although no published data are yet available on yields. It has the advantage of being owned and operated by a private company which has been responsible for the landfill operations of the city for many years. Being situated adjacent to the landfill site it enjoys a flexibility not available to many other mechanical sorting plants. A further advantage is that the company has extensive experience of materials recycling and the recovery markets and knows the quality specifications and quantity requirements of the secondary markets.

## Nancy, France

A major resource recovery facility started up earlier this year in Nancy. This is the first full-scale demonstration of the Revalord process developed by TRIGA and STRIVAL from experimental work performed by BRGM. The French Government through ANRED has assisted with the development costs.

The plant is designed to handle urban and industrial wastes and includes three parallel sorting lines with a total capacity of 22 t/hr. The composition of the waste inputs and the planned recovery levels of selected materials are shown in Tables C1.2(b) and C1.2(c).

199

| Table C1.2(b) | | | |
|---|---|---|---|
| PERCENTAGE COMPOSITION OF INCOMING WASTE | | | |
| Categories | Minimum | Maximum | Average |
| **+200 mm fraction** | | | |
| . cardboard | 3.0 | 8.0 | 4.4 |
| . newspapers, magazines | 4.0 | 10.0 | 6.6 |
| . plastic bags | 0.1 | 2.0 | 1.0 |
| . flexibles | 0.2 | 2.5 | 1.2 |
| . mix. materials | 0.2 | 2.0 | 1.0 |
| **-200 +50 mm fraction** | | | |
| . ferrous scrap | 1.5 | 6.0 | 3.3 |
| . non-ferrous metals | 0.1 | 0.5 | 0.3 |
| . glass | 4.0 | 10.0 | 7.0 |
| . PVC containers | 0.5 | 2.0 | 1.3 |
| . PE containers | 0.1 | 0.4 | 0.3 |
| . mix. light materials | 3.0 | 13.0 | 8.6 |
| . mix. organic materials | 28.0 | 42.0 | 34.0 |
| **- 50 mm fraction** | 17.0 | 37.0 | 31.0 |
| Total | | | 100.0 |

| Table C1.2(c) | | |
|---|---|---|
| PLANNED OUTPUTS OF VALUABLE MATERIALS | | |
| Recovered Products | Average Output/h | Recovery Efficiency |
| **+200 mm fraction** | | |
| . cardboards | 800 kg/h | 90% |
| . newspapers, magazines | 1,300 kg/h | 90% |
| **-200 +50 mm fraction** | | |
| . ferrous metals | 700 kg/h | 90% |
| . non-ferrous metals | 40 kg/h | 65% |
| . glass cullet | 1,300 kg/h | 85% |
| . PVC containers | 200 kg/h | 75% |

## Paper-fractions

The papers and cardboards recovered are greater than 200 mm and graded in two main categories:

o      the first is essentially cardboard boxes conforming to the European standard "corrugated board A5";

o      the second is essentially flat cardboards, magazines and newspapers complying with the European standard "mixed papers and cardboards no.2 A1".

Excluding structural contaminants (inks, staples, adhesives, etc.,) their impurities content (plastic, wood, rubber, leather, textiles) does not exceed 2% in weight.

The recovered materials are free of putrescible organic matters which could generate preservation problems, and their quality allows storage in bales for a minimum period of 3 months.

## Glass

The glass concentrate demonstrates the following characteristics, within glass manufacturing allowances:

o      size range

    •    upper dimension: 20 mm,

    •    lower dimension: fines content less than 5% in weight (fines is the name given to particles passing through a sieve the openings of which have a nominal size of 3 mm);

o      impurity content

The maximum quantities of contaminants do not exceed:

    •    0.1% of paper, plastic, wood, rubber,

    •    0.05% of stones, ceramics, other miner particles,

    •    0.01% of magnetic metals,

    •    0.05% of non-magnetic metals.

The mixed cullet is stored in containers.

## PVC bottles

The transparent bottles made of polyvinyl chloride are recovered with a grade compatible with the usual standards of the French market:

o      maximum content in paper: 5% in weight,

o        maximum content in polyethylene:  3% in weight,

o        maximum content in other impurities (glass, ceramics):
5% in weight.

They are compacted into bales ready to be converted into granules
for return to the regeneration industries.

<u>Ferrous metals</u>

The magnetic scrap sorted in the plant is essentially made up of
used tin cans, compressed into bundles for shipment to steel-
making mills or foundries.

<u>Aluminium</u>

The shredded aluminium concentrate meets target specifications
for reuse in cans and secondary alloy products.

It should be emphasised that the above are the design specifications; the plant has only been operational for around six months
and results are not yet available. It has been suggested that
initial operations have not gone smoothly and that the plant's
technical capability has yet to be proven. Similarly the ability
to market the recovered products at a price which gives an
economic performance has still to be established.

A second Revalord plant is being installed by the Syndicat de
l'Entre deux Mers but no information was available on the status
of this project.

**Other material recovery plants in Europe**

There are several other plants in Europe which have attempted
material recovery. These have either ceased operations or
changed their nature. For instance:

**VAM Recycling bV at Wijster, Holland**

This plant was designed to recover light paper, heavy paper,
organic material for composting, ferrous metals and
(subsequently) plastics. The plant was based on Fläkt
technology. The light paper was intended for the paper/board
mills. The specification requirement of 5% maximum contraries
has just about been achieved. However, in processing, the 5%
contraries became small in size compared to similar materials
found in mixed paper, thus making the removal of the 5% more
difficult. In the event neither the light or heavy papers found
a market outlet.

There were no technical recovery problems with the metals but the
market conditions were depressed due to the recession. Plastics
recovery was subsequently added with the intention of selling to
the Dutch State Mines. In the event matching the required
cleanliness of product, with high quantity recovery, was not
achieved. After 12 months of operation, markets had not been

found and a decision was taken to modify the plant to a basic separation system around several trommel stages. It is now reported that RDF is being considered.

**Enadimsa, Madrid**

A final plant which should be mentioned is the Enadimsa plant in Madrid. Originally based on technology developed by the US Bureau of Mines it has been completely redeveloped to suit Spanish conditions.

The plant is capable of separating ferrous metals, paper and plastics and also produced compost. At present the paper is diverted to the composting stream as there is no potential market closer than 350 km.

There are three categories of plastics which are sorted:

- light film plastic;

- hand sorted dense plastics;

- dirty reject.

The light plastic film is washed, shredded and baled for sale. The dense plastic, consisting mostly of plastic bottles used for oil, milk etc., is **hand sorted** into white (mainly polyethylene) and coloured plastic. It is then further shredded and bagged for sale.

There are three markets for the film plastic:

- extrusion and recycling;

- use with bitumen for road surfacing;

- solvent dissolution and recycling of the chemical product.

The latter outlet is still at the development stage and the main market appears to be extrusion and recycling.

The efficiencies for some of the separations are reported as being only 50% but reasonable product quality is being achieved.

C1.3    **Current Status**

For the material recovery plants which are currently operating no firm cost or economic data are available. It is therefore not possible to assess the true commercial nature of their operations. For instance it is not clear what level of subsidies are being provided by the local disposal authority to the plant operators. But it is clear from those plants which have not succeeded, e.g. Wijster, Vienna, and numerous plants in the United States) that it is difficult to integrate household waste into an economically viable circuit: refuse is composed of an infinite variety of products which have in common only that they are no longer wanted. The variability of the size and form of waste components, their uncertain composition, the presence of all types of contaminants and of pernicious contraries place a large demand of any mechanical sorting system which is required to produce a consistent product quality.

Because of these inherent problems it is likely that even if all the technical difficulties of mechanical recovery can be overcome the costs will be high. And how far these costs can be recovered will depend on success in marketing the products. This is rendered difficult for several reasons:

o        wastes tend to be available only in limited, variable and unpredictable quantities and qualities so that continuous, large-scale activity based on long term contracts is difficult to achieve;

o        certain specifications demand the absence of any reclaimed materials in the product;

o        social acceptance of reclaimed products is low;

o        the recovered material is usually of a lower grade than the virgin material which means it cannot act as a substitute and new outlets (products) have to be found;

o        cyclical trends in commodity markets make secondary materials very vulnerable at times of the downswing;

o        when a material becomes more valuable it will be less present in the waste either by substitution or by private separate collection reducing the value of the refuse.

Given these uncertainties total dependence on a material recovery plant for disposal has a high risk element. The inhomogeneity of the feed and changing market conditions mean that flexibility must be built into the disposal arrangements. Either alternative disposal route needs to be available such as at Neuss, for instance, where the plant is located on a landfill site or processes have to be developed which can easily be modified to derive other products, in response to market values. Thus the ability to switch combustibles (paper and plastics) between material recovery on the one hand and RDF on the other could be an advantage. Short term switches of this sort however do have disadvantages in that long term commitments for any substantial proportion of the output cannot be made. There is also the question of the higher investment required. The greater the flexibility built into the plant the higher the cost.

Even if a plant is built with the capability of responding rapidly to market conditions it will only do so if the plant's management react in the appropriate manner. This requires a more entrepreneurial approach on the part of waste disposal authorities than is normally associated with such bodies. It is necessary to stay in close contact with the secondary material markets, anticipate trends and respond rapidly. Given the inevitable difficulties for public authorities to act in this manner there are advantages in contracting out responsibility for the plant to private operators.

**C1.4        The Future Potential of Mechanical Separation**

**C1.4.1     General**

Mechanical recovery techniques are still in their infancy. It is unlikely that they will ever provide a universal panacea to waste disposal and material recovery. Even if other technical, economic and market difficulties can be overcome, they will only be worthwhile considering in locations where a sufficient volume of wastes can be collected and brought to a centralised site. This limits their potential to urban centres; the mechanical recovery of dispersed wastes in rural areas would be hopelessly uneconomic.

It is quite likely that commercial separation plants will be installed in urban areas in future years but only on a highly selective basis. The ingredients for such schemes might be:

o        private operator with experience of secondary markets;

o        alternative disposal options for when market conditions do not suit recovery;

o        maximum flexibility in plant to switch recovery between materials and materials and RDF;

o        a tailor-made design to suit specific conditions relating to waste composition and product markets.

**C1.4.2     Potential by material**

**Paper recovery:** Much of the early work on mechanical separation has been devoted to producing a secondary fibre of sufficient quality to be reused in paper and board production. This has been achieved at a basic level and work is continuing of developing classiciation and sorting systems to provide a better quality product for wider application in the paper industry. There has been considerable difficulty in finding markets for mechanically separated paper.

**Aluminium recovery:** The technology exists for the mechanical recovery of aluminium from municipal refuse but has not so far been adopted on a commercial scale in Europe. This is partly because in many countries aluminium constitutes a very small proportion of the waste and economic recovery has not been possible.

**Plastics recovery:** The recovery of plastic film is undertaken in Sorain-Cecchini plants in Rome. In addition to film, where plastic bottles are a significant constituent of household wastes these can be separated as a product. There is still considerable development required on the separation and cleaning of plastics concentrates. In particular until satisfactory separation of mixed polymer products can be achieved, their recycling will be limited to applications where such mixtures are acceptable.

**Glass recovery:** The technology for glass (cullet) recovery (from flotation and optical sorting) is also available although it has not been widely applied to date.

**In general,** a major limitation of mechanical recovery systems so far has been the low quality of the reclaimed materials. This limits marketability and while basic technology exists there is a general need to improve this to produce higher specification recovered materials.

It is unlikely that a recovery system devoted to a single material will ever be viable. The above materials are thus likely to be recovered in combination either with each other or with RDF. Ferrous metals will also certainly be recovered at the same time. This in turn has implications for the size of the capital investment required in such plants. The processes used in recovery plants require a minimum throughput for economic operation.

**C2**    <u>Separation at Source</u>

The source separation of materials is a well established method of recycling materials. This can take many forms but there are two basic approaches which can be adopted:

o         separated at source (home, office, etc.) for collection by a third party;

o         separated at source and taken to a local or centrally located container for collection by a third party.

There are many variations on each of these schemes and below we consider the main activities in this area for glass, aluminium, plastics and paper recovery in Europe.

**C2.1**    **Paper**

The recovery of paper from household and trade sources has long been established. As already noted any additional supplies of waste paper will be derived from household waste. The collection of wastes directly from households are at present undertaken by Member Countries either by the local authorities or volunatry organisations (such as schools, clubs, the Red Cross, etc.).

There has been much discussion on the cyclical nature of the waste paper market and the role of local authorities within it.

Because of the cyclical instability of the waste paper market, local authorities have been reluctant to invest in the necessary equipment or establish the necessary organisational framework in the face of periodic collapses in waste paper requirements and prices. Various solutions have been suggested including intervention by public authorities to establish a guaranteed price for waste paper or, alternatively, the arrangement of long-term contracts.

In contrast to voluntary organisation collection activities, waste paper recovery by local authorities is not so readily turned on and off intermittently, especially if capital investment has been undertaken. Contractual arrangements between local authorities and merchants/mills for the supply of minimum quantities of waste paper at specified prices would provide a more stable basis for waste paper recovery.

The fact remains that local authority involvement in waste paper collection is patchy and something less than wholehearted. Thus despite encouragement from national governments throughout the Community, while cooperation from households is likely to be achieved up to reasonable levels the problem of disposing of relatively low grade paper remains. It is a problem of demand as much as supply. There is also the fact that where local authorities operate incineration plants there is likely to be a reluctance to divert a major combustible material from this disposal route. This is particularly the case where incineration is associated with heat recovery or electricity generation. In

the Netherlands, for instance, a considerable proportion of
domestic waste is incinerated in four installations which
generate energy.  A waste supply of lower calorific value could
reduce revenues from electricity sales more than variable cost
savings of disposal.

C2.2     **Glass**

This is the other major material arising in domestic refuse for
which a history of separate collection has been established.  In
1982, glass recycling in Europe exceeded two million tonnes.  The
volumes recovered and the percentages this represents for
different countries in the EEC is shown below.

| Table C2.2(a) | | |
|---|---|---|
| EUROPEAN GLASS RECYCLING - 1982 | | |
| Country | Tonnes* Recycled | Proportion of National Glass Consumption |
| Belgium | 100,000 | 32% |
| Denmark | 21,500 | 10% |
| France | 478,000 | 20% |
| Germany | 750,000 | 28% |
| Great Britain | 110,000 | 6% |
| Ireland | 6,600 | 8% |
| Italy | 355,500 | 21% |
| Netherlands | 200,000 | 47% |
| Total | 2,021,600 | |
| *Excludes manufacturers cullet. | | |

Unlike paper, which has traditionally been collected on a door-
to-door basis, glass recycling relies much more heavily on householders taking glass to a central collection point (bottle bank
schemes).  Door-to-door collection is still practised, either as
a separate collection or in combination with other materials
(e.g. paper and/or plastic), but bottle banks are an increasingly
common feature of glass recovery schemes.

**Bottle banks:**  Typically large skips, usually containing three
compartments, are placed in prominent locations; the public
deposits waste bottles into the skip using the appropriate
compartment for different colours of glass.  At regular
intervals, or when the skips are full, the local authority
transports them to a central storage area for emptying and then
returns them to the collection site.  When a sufficient quantity
of glass of any one colour has accumulated at the central storage
area it is sent to a cullet processing plant.

A modification of this scheme (as being introduced in the UK) is to use smaller bins, each for a single colour, which allows far more flexibility in location. Also they can be emptied at the collection location since most are designed for operation with special emptying vehicles.

Colour separation

The major part of glass container production in Europe is devoted to three colours: clear, green and brown. Ideally, when recycling, the colours should be kept separate and returned for remelting in the same colour of furnace from which they came. If the colours become mixed, the last resort is to put them all into a furnace producing green glass.

However, collecting separate colours nearly always makes it more expensive - extra skips, more complex vehicles, larger storage bays - and, initially, many schemes were set up based on mixed collection. While there is sufficient capacity in the green furnaces no problems arise, but as glass recycling grows there comes a point when the green furnaces reach saturation. Already in some countries they are operating at 85% cullet or more and it is evident that the true potential for glass recycling cannot be reached without colour separation.

Because it has virtually no wine production and, by European standards, a low consumption of wine, the UK glass industry makes a low proportion of green bottles - approximately 13%. Thus, a system of mixed collection would limit UK glass recycling to about 10% of national consumption, whereas many other countries have already passed 30%.

At the other end of the scale, some 50% of French glass container production is green, and thus the achievement of 20% recycling is possible even without colour separation. Nonetheless the glass producers in France are studying ways to introduce schemes for colour separation in parts of the country.

West Germany already operates separate collection of colours in many areas. Indeed, without this the recycling rate of 28% would have been impossible because green glass manufacture is about 29% of the Federal Republic's output. Clear glass presently makes up some 47% of production, whilst clear cullet is just under 16% of collections.

Some balancing of supply and demand between countries is possible but the value of cullet does not encourage export over long distances and, in any case, all countries face a similar problem in the longer term.

Two methods have been tried to separate colours after collection:

o       by manual sorting - not an easy or economic task;

o       by mechanical devices based on a 'magic eye' - equally expensive and difficult to operate in practice.

For the present therefore, it seems that to ask the general
public to put the colours into different holes or containers is
the most practical and economic solution.

Because the raw materials for clear glass are often more
expensive than for coloured, glass manufacturers are usually able
to offer higher prices for cullet separated into colours.

C2.3        **Aluminium**

The main recovery schemes so far instituted for aluminium can
recovery differ in one important respect from the schemes used
for paper and glass recovery in that a direct financial incentive
is offered to the householder.

Financially motivated aluminium can recycling has been operating
in the United States for at least fifteen years.  There are over
2,000 buying locations around the country, including several
hundred mobile units which operate from central and satellite
centres on an established, advertised schedule, buying aluminium
scrap from the public.  In 1980, 600 m pounds were recycled
domestically through the consumer recycling network.  It is
estimated that around 30% of all aluminium cans are recovered
countrywide with rates as high as 60% in some areas.

The inherent value of the metal enables a price to be paid for
the cans while still maintaining economic viability.  Consumers
take the cans to a site where they are weighed and exchanged for
cash on a weight basis (after magnetic separation of steel cans).
The aluminium cans are then taken to a central treatment area
where they are either shredded or baled before dispatch, usually
by rail, to the aluminium smelter.  The can scrap is suitable for
use directly back into cans or for similar alloy applications.

The penetration of all aluminium cans in the European beverages
market is very much lower than in the United States, although
penetration has increased in recent years.  Examples of aluminium
can shares in Europe are as follows:

| | | |
|---|---|---|
| Germany | : | 10% |
| UK | : | 45% |
| Italy | : | 50% |
| France | : | negligible |

It can be seen that in the EEC, Italy and the UK have the highest
penetration rates and recovery schemes are now being established
in these countries.  In the UK, Alcoa have set up schemes based
on the US model whereby cans are purchased at a number of centres
throughout the UK.  So far schemes have been established in seven
cities; anyone taking cans to a cash-a-can processing centre
receives direct payment for their cans (curently 40p per kilo -
approximately 1p per can).  If the cans are taken to a mobile
collection site or if the cans are collected from a collector's

location, this reduces to 0.5p per can.  In 1982, around 14 million aluminium cans were collected; total UK consumption is around 1.45 billion.  In the largest established centres recovery rates of around 12% have been attained.

For the scheme to become economically viable it has been estimated that aluminium cans will need to take a 70% share of the canned beverages market and a recovery rate of 30% obtained.

To save consumers having to travel to processing centres or collection points, collection through machines located in super- markets is being tested.  These are essentially reverse vending machines which receive, check, crush and store the cans and issue receipts or cash in return.  Whether supermarkets will be willing to sacrifice space in their stores to locate this equipment remains to be seen.

**C2.4     Plastics**

The recovery of plastic bottles (PVC OR PET) is feasible by any of the methods so far considered.  In the case of deposit bottles the return of these to stores for recyclying to bottlers is a standard practice.

In the case of non-deposit bottles, voluntary recycling through central collection points is being tested.  In the UK, at Bradford and Leeds, PET-a-Box skips have been installed at central locations.

**C2.5     Summary**

It is beyond the scope of this study to make a detailed examina- tion of alternative source separation schemes.  Many different types of schemes have been tried throughout the Community based both on door-to-door collection and consumers' taking to central collection points.  The latter has advantages in cost terms but disadvantages in energy terms.  Both types of schemes are dependent upon consumer participation which has been highly variable between countries in the Community and between product groups.  But there are no technical difficulties to be overcome and if consumer cooperation can be obtained, high levels of recovery are feasible.

Appendix D

SYSTEMS FOR THERMAL CONVERSION BY PYROLYSIS

**D1**   **PYROLYSIS**

**Definition**

Pyrolysis is the thermal degradation of materials in the absence of oxygen in order to produce hydrocarbon rich oils or gases and char from organic materials.

Figure D.1(a) illustrates the flow diagrams for the major pyrolysis systems detailed below.

**D2**   **PYROLYSIS OF PLASTICS**

**D2.1**   General

As with direct incineration no waste pretreatment is necessary, although rough shredding to reduce size of plastics material to about 60 mm would be advantageous and increase efficiency of conversion. Mixed wastes can be treated in a similar manner to pure waste streams, e.g. used cable pyrolysis and both batch and continuous processes exist.

There are a number of major drawbacks however; pyrolysis is a relatively new technology and not totally proven and commercially produced units are only just entering the waste treatment market. The principal advantages of pyrolysis systems are that they produce a fuel which can be stored and used as and when necessary and that generally they operate without major environmental impact.

Major types of pyrolysis units for plastics wastes are:

o        rotating and non-rotating kilns;

o        molten media reactor;

o        fluidised-bed reactor.

**D2.2**   State of technology

**Non-rotating kilns** have been applied to used cables with external heating but have encountered severe problems with heat transfer efficiency to the plastic wastes.

Both Mitsubishi Heavy Industries and Mitsui Shipbuilding and Engineering Company have experimented with **molten media** reactors for plastics. The molten plastics are pumped into an externally heated reactor which is agitated to maintain heat transfer and prevent charring. High molecular weight materials, essentially tars, are removed from the bottom of the vessel and oils liberated are condensed in a reflux condenser.

FIGURE D.1(a):   WASTE PYROLYSIS SYSTEMS

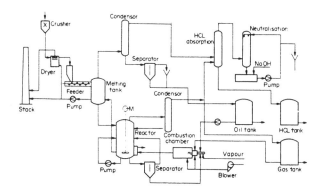

A.   Molten-bed pyrolysis unit

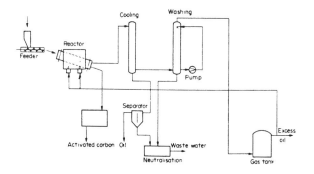

B. Potary-kiln pyrolysis unit

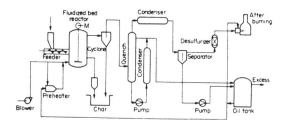

C. Fluidised bed pyrolysis unit

Research into the pyrolysis of plastics in a **fluidised-bed unit**
at the University of Hamburg under Kaminsky has been particularly
promising. The programme has been funded in part by both the
Federal Ministry for Research and Development (BMFT) and the
German Association of Plastics Manufacturers (VKE). In contrast
to research work in Japan, the German project is using relatively
higher temperatures (600-900°C) and produces an aromatic oil.
Fluidised bed units are seen to offer distinct advantages over
other pyrolysis systems, in that:

o        there are no mechanical or moving parts in the hot area;

o        homogeneous temperature fields and good heat distribu-
         tion are attainable;

o        a more reliable and standardised pyrolysis product
         results;

o        the system can be completely enclosed and requires no
         complicated kiln scaling systems as it is fixed;

o        solids carried out of the reactor zone can be easily
         separated by a cyclone meeting air emission regulations.

Currently a laboratory plant capable of firing up to 20 kg/h of
polyethylene and polypropylene has been constructed and it is
intended to progress ultimately to a 100 kg/h pilot plant.

Continued research is underway in Germany and Japan on **rotary
kiln** pyrolysis. Principal developments are the:

o        Kiener Process

o        Babcock Kraus Maffei Process

o        Mannesman Veba Umwelttechnik 'Rotopyr' Process.

All three systems are being developed for the pyrolysis of mixed
wastes; Kiener & Babcock for municipal wastes and Mannesman for
industrial wastes although pyrolysis of plastics has been extensively experimented with at the test units owned by each
developer.

The systems differ in their method of heat applications and
control of air emissions; Kiener utilises dual stage alkali/acid
scrubbing of the wastes' gases whereas Babcock introduces time to
neutralise acidic components generated. Both Babcock  and
Mannesman apply heat to the pyrolysis kiln externally whereas the
Kiener system utilises internal hot gas pipes. All three systems
use the pyrolysis gas generated for heating the vessel following
removal of the condensable oil products formed. The objective is
to produce oils which could be substituted for plastic feedstock.

The Babcock system is currently undergoing commissioning trials for a 6 t/h full-scale plant for the town of Ulm in the Landkreiss of Günzburg. The Kiener system is shortly to progress to a 6 t/h (100%) scale-up plant at Aalen following extensive testing of a single time 3 t/h plant. Both systems are however using domestic and bulky wastes as their substate. The Rotopyr system has been experimented with on a 200 kg/h scale plant at Bochum using a variety of substrates including waste mixed plastics and redundant cable.

D2.3    **Energy yield**

Both the Kiener and Babcock units have achieved high levels of availability and energy efficiency at the pilot scale. On the basis of this experience it would appear reasonable to adopt their conversion efficiencies of between 50 and 60% net.

D3      **PYROLYSIS OF RUBBER**

D3.1    **General**

As with plastics there has been a considerable number of developments in pyrolysis of rubber wastes in Germany and elsewhere during the last five years.

Both Kiener and Babcock have experimented with rubber wastes on their respective test units and Kiener have incoporated rubber wastes in their municipal waste unit at Aalen. More significant, however, is the unit constructed at Reutlingen following the trials conducted by the University of Hamburg.

D3.2    **State of Technology**

   i)    **European experience**

The University of Hamburg system is a **fluidised-bed unit,** operating at 2 tonnes per hour and capable of firing tyre material up to 2.7 kg in weight, giving large savings in the cost of shredding. The fluidised-bed region is heated indirectly by a series of seven radiant fine tubes, the gas produced by the pyrolysis is used both for fluidising and heating the sand bed. In addition to producing the pyrolysis gases and oils, a quantity of carbon black (35-45% W/W of the total product's mass) is produced. The project is funded by BMFT (the German Ministry of Research and Development) and the engineering company Carl Robert Eckleman.

No problems are expected in the operation of the Reutlingen plant except that such plants for the thermal conversion of rubber wastes in Germany have historically suffered from major difficulties in securing consistent and adequate supplies.

In France the Compagnie General de Chauffre intends to use scrap tyres for district heating following conversion by the **Pyralox process**. A further Pyralox unit is owned by the Michelin Tyre Company subsidiary Prier Lauvent to produce steam for heating in a retreading factory. These developments have occurred in spite of the price of competitive fuels such as coke, but the French Government Agency predicts that future developments will only be stimulated if public subsidies are provided and collections of scrap tyres guaranteed. Additionally pyrolysis is capital intensive and can only be truly successful at higher scales of operation.

Connected depolymerisation experiments have been carried out in France under the ANRED sponsored project aimed at depolymerisation of rubber. This process involves the addition of scrap tyres to an oil bath maintained at a temperature of about 90-95°C to produce a fuel-oil substitute. The project is now undergoing scale-up and although successful technically it is anticipated that the process may ultimately prove too costly both in terms of capital and operational costs. The process would utilise 12,000 tonnes of oil per annum to treat 4,000 tonnes of tyre giving 15,200 tonnes of combustible oil and 800 tonnes of metals and textiles.

Fiat have been cooperating with the American company Intenco since 1978 on the development of a tyre pyrolysis process which could progress to installation of three regional processing centres. The 100 tonnes per day plants will produce fuel oil, carbon black and scrap steel, any gases generated are recirculated to maintain the process heat. In 1980 negotiations were underway to construct a plant between Turin and Milan but problems were anticipated in securing suitable outlets for the carbon black generated.

### ii)    Other experience

Several developments have occurred in America, namely the pilot studies of USBM/Firestone and the Goodyear/Tosco projects, but many of these proved uneconomic and proceeded no further. Foster Wheeler in the UK, however, have developed the cross-flow pyrolysis process designed by Warren Spring and prepared a 100 tonnes per day plant. This plant is to be operated by a subsidiary of Foster Wheeler from the end of 1983. The unit will process 50,000 tonnes of tyres per annum yielding 20,000 tonnes of light fuel oil, 17,000 tonnes of carbon black and 7,000 tonnes of steel scrap. Figure D3.2(a) comprises a flow diagram for the process.

Developments in Japan, e.g. the Hyogo Industrial Group, a joint government/private industry venture, have been particularly concerned with the production of carbon black of consistent and marketable quality from scrap tyres. This necessitates intensive mechanical processing to produce very fine crumb of 5 um which can then be pyrolysed. This, therefore, makes the process both uneconomic and reduces the energy yield of the system considerably

FIGURE D3.2(a):   TYROLYSIS LTD. (FOSTER WHEELER POWER PRODUCTS LTD.)

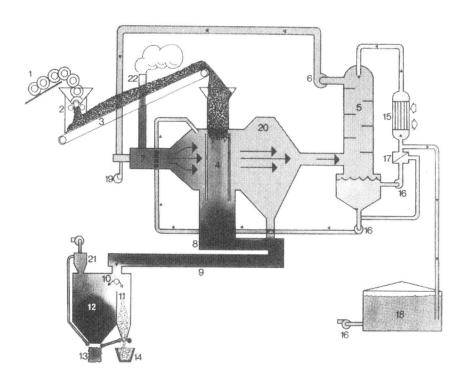

## Tyre Pyrolysis Unit

1. Conveyor
2. Shredder
3. Conveyor
4. Pyrolyser
5. Quench Tower
6. Re-circ. fan
7. Furnace
8. Conveyor
9. Conveyor
10. Magnetic separator
11. Steel hopper

12. Char hopper
13. Bagger
14. Skip
15. Cooler
16. Pumps
17. Filter
18. Storage tank
19. Burner air fan
20. Expansion chamber
21. Char dust collector
22. Stack

## Manufacturer's Specification

100 tons of scrap tyres typically produces:

40 tons of oil; 35 tons of char; 10 tons of steel wire (dependent upon type and size of tyre); 15 tons of fuel gases

The Kobe system utilises a rotary kiln to process 5 tonnes/day of tyres to produce pyrolysis oils, gases and char at a temperature of about 850°C.

Due to the large number of potentially commercially viable rubber waste pyrolysis units there is a great reluctance to publish any data on process economics.

## D3.3     Energy Yield

Energy yields are likely to be equivalent to those for municipal wastes.

## D4     PYROLYSIS OF WOOD

## D4.1     General

Development of pyrolysis and gasification systems for wood wastes was widespread 40 years ago. Pyrolysis is not an adopted technology for wood wastes alone, although the Germany company Fritz Werner GmbH have developed both medium- and large-scale units for conversion of wood to charcoal or oils for tropical countries.

Fritz Werner have also been prominent in developing a series of wood gasification plants for similar markets, particularly developing countries. Considerable research continues both in Europe on gasification of agricultural wastes and in South America on wood.

## D4.2     State of Technology

Several systems have been developed and these are essentially either:

o      fixed-bed gasifiers, or

o      fluidised-bed gasifiers.

Fixed-bed gasifiers are an established technology and commonly produce high calorific value gases low in tar, phenol, $CO_2$ and nitrogen and these are commonly linked, at small scale, to reservoirs for vehicle fuelling or producing electrical energy by converted diesel engines.

Fluidised-bed gasifier development occurs under Fritz Werner in Germany for materials of variable particle size. The bed operates at a temperature of 1,000°C at which all hydrocarbons present are cracked. Feeding of wood chips or bark is accomplished by mechanical means due to bridging difficulties on storage and transfer. Typical energy conversion efficiencies claimed are approximately 75%.

**D4.3**    **Energy Yield**

The Fritz Werner gasification system for wood is between 65 and 70% thermally efficient which is slightly higher than that for mixed municipal wastes achieved by Kiener and Babcock Kraus Maffei.

221

Appendix E

MODIFICATIONS TO ENERGY SAVINGS
DUE TO THE IMPORT OF MATERIAL
(raw, semi-processed and processed)
FROM OUTSIDE EUR 10

## E1.    INTRODUCTION

In calculating the energy savings which accrue to EUR 10 through increased secondary production, it is necessary to deduct the energy savings which occur outside EUR 10. The energy element of materials and products which are incurred externally do not constitute energy savings to EUR 10 when displaced by increased secondary production within EUR 10.

It is necessary therefore to apply an adjustment factor to the total energy savings figure to derive energy savings specific to EUR 10. This has been calculated for each material separately, as described below.

## E2.    ALUMINIUM

In addition to the 2.21 Mt of aluminium produced in the EUR 10 in 1981, of which 1.17 Mt were produced by the secondary smelting of scrap, 0.95 Mt and 0.27 Mt respectively were imported as raw ingot/alloy and finished products (bar, wire, plate, sheet, strip, foil, powder/flakes and tube). There are negligible imports of scrap aluminium.

Uses of imported and home produced aluminium are generally interchangeable although there are exceptions, such as high purity requirements for wire drawing etc. Recycled aluminium can therefore be smelted and substituted for primary aluminium (traditionally in the forgings sector) and can be assumed to displace a fraction of primary aluminium supplies. This is assumed to occur on a pro-rata basis between imports and home production.

The calculation of the effect of extra home secondary production on the energy requirements of the EUR 10 can therefore be calculated on the assumption that all secondary material physically available could be resmelted.

Thus:

| | | |
|---|---|---|
| Home production | | |
| $H_p$, primary: | 1.04 Mt |
| $H_s$, secondary: | 1.17 Mt |
| | | |
| Import | | |
| $I_i$, ingot/alloy: | 0.95 Mt |
| $I_p$, products: | 0.27 Mt |

$R_p$, further recovery potential (physically available) = 0.31 Mt

Assuming relative energy values of materials as follows:

| | | |
|---|---|---|
| $E_{pi}$, primary ingot production | = 1.00 |
| $E_{pm}$, primary material production | = 1.10 |
| $E_{sm}$, secondary materials production from scrap | = 0.11 |

i.e. conversion of bauxite in the ground to aluminium materials (section, bar, rod, etc.) requires 10% more energy when carried out directly than for ingot production alone. Furthermore, secondary smelting and forming requires only 10% of the energy expended in primary production of materials from bauxite in the ground.

The significance of the energy content of the imported materials can be calculated knowing the relative displacement of primary home produced and imported materials, which is:

|  | 1981 Data (Mt) | Displaced by Fraction of $R_p$ | Corrected Production Levels to Account for $R_p$ | Denoted |
|---|---|---|---|---|
| $H_p$ | 1.04 | 0.14 | 0.90 | $H_{p2}$ |
| $I_i$ | 0.95 | 0.13 | 0.82 | $I_{i2}$ |
| $I_p$ | 0.27 | 0.04 | 0.23 | $I_{p2}$ |
| Total | 2.26 | 0.31 | 1.95 | |

Therefore energy content can be calculated as:

$$= \frac{I_{i2} \times E_{pi} + (I_{p2} \times E_{pm})}{(R_p \times E_{sm}) + (H_s \times E_{sm}) + (H_{p2} \times E_{pi}) + (I_{i2} \times E_{pi}) + (I_{p2} \times E_{pm})}$$

$$= \frac{0.82 + (0.23 \times 1.10)}{(0.31 \times 0.11) + (1.17 \times 0.11) + 0.90 + 0.82 + (0.23 \times 1.10)}$$

$$= \frac{1.073}{2.136} \qquad = 0.502$$

Thus, the total potential energy savings which could be achieved by the recycling of an extra 0.31 Mt must be reduced by 50.2%. This is the proportion of energy which is not saved within EUR 10 when aluminium is replaced by increased secondary home production.

**E3.** **PLASTICS**

In addition to the 20 Mt produced in the EUR 10 in 1981, of which
1.64 Mt were produced from secondary sources (excluding imported
plastic scrap), 1.5 Mt were imported into the EUR 10 from third
countries. This 1.5 Mt can be broken down further into the 1.18
Mt imported as primary goods or process raw materials and 0.30 Mt
imported as finished articles e.g. shapes, film, strip, tube,
etc.  0.07 Mt were imported to the EUR 10 as waste plastics.

Assuming an approximately equal split between shaped products and
films this gives an average conversion energy from polymer resin
to products of around 25 GJ/t (shapes 10GJ/t, films 40 GJ/t).
Further assumptions are a mean primary energy consumption of 120
GJ/t for polymer provision (92.32 GJ/t from crude oil, to 222.14
GJ/t for polyester resins), and a conversion energy of 26 GJ/t
for secondary materials (including 0.59 GJ/t for sorting and 0.41
GJ/t for cleaning raw material), to new products.

This allows calculation of the total potential energy savings
modifier:

Home production:

$H_p$, primary      = 18.36 Mt
$H_s$, secondary    =  1.64 Mt

Import

$I_r$, polymer resin  = 1.18 Mt
$I_p$, products       = 0.30 Mt
$I_w$, waste plastics  = 0.07 Mt
$R_p$, further recovery potential (physically available) = 10.08 Mt

Assume relative energy values of materials as follows:

$E_{pr}$, primary energy assumption of resin production = 1.00
$E_{pp}$, primary production of products                = 1.21
$E_{sp}$, secondary production of products from scrap   = 0.22

i.e. conversion of plastics resin to products requires a further
21% of the primary energy  requirement to furnish the resin
whereas remelting and reforming consumes only 22% of the original
primary requirement.

Calculation of market displacement:

| | 1981 data | Displaced by fraction of $R_p$ | Corrected production and import levels | Denoted |
|---|---|---|---|---|
| $H_p$ | 18.36 | 9.29 | 9.07 | $H_{p2}$ |
| $I_r$ | 1.18 | 0.60 | 0.58 | $I_{r2}$ |
| $I_p$ | 0.30 | 0.15 | 0.15 | $I_{p2}$ |
| $I_w$ | 0.07 | 0.04 | 0.03 | $I_{w2}$ |
| Total | 19.91 | 10.08 | 9.83 | - |

Relative energy content of imported material can be calculated as:

$$\frac{(I_{r2} \times E_{pr})+(I_{p2} \times E_{pp})+(I_{w2} \times Epr)}{(R_p \times E_{sp})+(H_{p2} \times E_{pp})+ (H_s \times E_{sp})+(I_{r2} \times E_{pr})+(I_{p2} \times E_{pp}) + (I_{w2} \times E_{pr})}$$

$$= \frac{(0.58)+(0.15 \times 1.21) + 0.03}{(10.08 \times 0.22) + 9.07 + (1.64 \times 0.22) + 0.58 (0.15 \times 1.21) + 0.01}$$

$$= \frac{0.79}{12.44} \qquad = 0.064$$

i.e. the total potential energy savings which could be achieved by the recycling of an extra 10.08 Mt must be reduced by 6.4% as 6.4% of the energy involved was initially expanded extra EUR 10 in providing:

o        primary polymer resins
o        finished products
o        waste plastics for conversion intra-EUR 10

## E4.    PAPER AND BOARD

In addition to the 25,320 Mt of paper and board produced in EUR 10 in 1981, of which about 12.00 Mt came from secondary production, 0.56 Mt and 8.59 Mt were imported as waste paper and pulp respectively from third countries.  In addition, 8.15 Mt of all paper grades were imported, including 3.46 Mt of newsprint. Home production of wood pulp in 1981 intra-EUR 10 was 5.56 Mt.

The market and material substitutions afforded by the different grades of waste paper and pulp are complex.  For the purpose of this analysis of relative extra- and intra-EUR 10 energy contributions we have assumed that all materials are directly interchangeable in paper and board preparation.  Furthermore it is assumed that increased levels of home recovery will occur at the expense of home primary production and importations on a pro-rata basis (this includes pulp for paper production) and ignores material losses on conversion of pulp to paper).

Calculation of the effect of this increased level of home production on the energy requirements of the EUR 10  can therefore be made on the basis that all secondary resources physically available are directed for recovery.

Thus:

Home production

$H_p$, home primary paper production    = 13.32 Mt
$H_s$, home secondary paper production = 12.00 Mt
$H_{pp}$, home primary pulp production    =  5.56 Mt

Import

$I_p$, imported paper production     =  8.15 Mt
$I_w$, imported waste paper          =  0.56 Mt
$I_{pp}$, imported paper pulp          =  8.59 Mt

$R_p$, further recovery potential (physically available)
                                     =  16.96 Mt

Assume relative energy values of materials as follows:

$E_{pp}$, energy of primary production                  = 1.00
$E_{pm}$, energy of primary paper and board production   = 1.92
$E_{sm}$, energy of secondary paper and board production = 0.54

Thus the conversion of wood pulp to paper in an integrated paper factory requires 92% more energy than the provision of pulp from standing timber while the conversion of waste paper to paper requires only 54% of the energy required to manufacture primary pulp from standing timber.

The significance of the energy content of the imported materials can be calculated assuming the relative displacement of primary home produced and assorted imported materials as follows:

| | 1981 data | Displaced by fraction of $R_p$ | Corrected production and import levels to account for $R_p$ (Mt) | Denoted |
|---|---|---|---|---|
| $H_p$ | 13.32 | 6.25 | 7.07 | $H_{p2}$ |
| $H_{pp}$ | 5.56 | 2.61 | 2.95 | $H_{pp2}$ |
| $I_p$ | 8.15 | 3.82 | 4.33 | $I_{p2}$ |
| $I_w$ | 0.56 | 0.26 | 0.30 | $I_{w2}$ |
| $I_{pp}$ | 8.59 | 4.03 | 4.56 | $I_{pp2}$ |
| Total | 36.18 | 16.96 | 19.21 | - |

Thus the relative energy content of imported material can be calculated as:

$$\frac{(I_{p2} \times E_{pm})+(I_{w2} \times E_{pm})+(I_{pp2} \times E_{pp})}{(R_p \times E_{sm}) \times H_{p2} \times E_{pm})+(H_s \times E_{sm})+(H_{pp2} \times E_{pp})+(I_{p2} \times E_{pm})+(I_{w2} \times E_{pm})+(I_{pp2} \times E_{pp})}$$

=

$$\frac{(4.33 \times 1.92) + (0.30 \times 1.92) + 4.56}{(16.96 \times 0.54)+(7.07 \times 1.92)+(12.00 \times 0.54)+2.95+(4.33 \times 1.92)+(0.30 \times 1.92) + 4.56}$$

$$= \frac{13.45}{45.61} = 0.29$$

i.e. the total potential energy savings which could be achieved by the recycling of an extra 16.96 Mt of waste paper must be reduced by 29%. This is the proportion of the total energy involved in the provision of materials which would be expended extra-EUR 10, assuming increased home secondary production displaces proportionate levels of home and imported materials.

**E5.**  **GLASS**

In addition to the home production of glass in the EUR 10 of 12.643 Mt in 1981 [1], of which about 1.80 Mt came from indigenous secondary materials recovery, 0.03 Mt of cullet and 0.90 Mt of all grades of glass respectively were imported from third countries extra-EUR 10.

The market and material substitutions for primary produced glass by cullet are well defined. Currently cullet substitution from prompt wastes is between 10 and 20% and substitution to a level of 80% or more may be acceptable within certain glass grades, principally green or brown hollow container glass. In this analysis we have assumed that a ready market would exist for all recoverable glass as direct substitution for imported cullet and glass and home produced primary glass.

Home Production

| | |
|---|---|
| $H_p$, home primary glass production | 10.84 Mt |
| $H_s$, home secondary production | 1.80 Mt |

Imports

| | |
|---|---|
| $I_g$, imported glass products | 0.90 Mt |
| $I_c$, imported cullet | 0.03 Mt |

$R_p$, further recovery potential (physically available) = 7.10 Mt

Assuming relative energy values of materials as follows:

| | |
|---|---|
| $E_p$, energy of primary glass production | 1.00 |
| $E_s$, energy of secondary glass production | 0.75 |

i.e. at a 100% substitution in melt, or glass production from cullet alone, a 25% energy saving is achieved.

The significance of the energy content of imported materials can be calculated knowing the relative displacement of home produced and imported glass and imported cullet, which is:

---

[1] Estimate of production according to European Glass Makers Federation (FEVE) 1981 was 15,880 Mt. The data utilised were provided by Eurostat, Industrial Statistics Bulletins.

| | 1981 data | Displaced by fraction of $R_p$ (Mt) | Corrected production and import levels to account for $R_p$ (Mt) | Denoted |
|---|---|---|---|---|
| $H_p$ | 10.84 | 5.67 | 5.17 | $H_{p2}$ |
| $H_s$ | 1.80 | 0.94 | 0.86 | $H_{s2}$ |
| $I_g$ | 0.90 | 0.47 | 0.43 | $I_{g2}$ |
| $I_c$ | 0.03 | 0.02 | 0.01 | $I_{c2}$ |
| Total | 13.57 | 7.10 | 6.47 | - |

Relative energy content of imported materials can be calculated as:

$$= \frac{(I_{g2} \times E_p) + (I_{c2} + E_p)}{(R_p \times E_s) + (H_{p2} \times E_p) + (H_s \times E_s) + (I_{g2} \times E_p) + (I_{c2} \times E_p)}$$

$$= \frac{0.43 \times 0.01}{(7.10 \times 0.75) + (5.17) + (1.80 \times 0.75) + 0.43 + 0.01}$$

$$= \frac{0.44}{12.29} = 0.036$$

i.e. the total potential energy savings which could be achieved by the recycling of an extra 7.10 Mt of waste glass intra EUR 10 must be reduced by 3.6%. This is the proportion of the energy initially expended extra-EUR 10 in providing:

o            glass products
o            cullet for import into the EUR 10.

which is replaced by secondary production.

**E6.**     **RUBBER**

In addition to the home production of approximately 20 Mt in 1981, of which

o        1.63 Mt was synthetic rubbers and

o        0.50 Mt was secondary rubber production,

0.968 Mt was imported as raw materials, 0.029 Mt as rubber scrap and 0.296 Mt as rubber products of which 46% was natural rubber material, the remainder being synthetic polymers such as SBR and IIR.

The market and material substitution possibilities for rubber are confined at present due to:

o        cheap imports;

o        changing composition and specifications for products;

o        insurgence of plastic and other synthetic products into traditional rubber markets, e.g. shoe soles.

For this analysis, however, we have assumed that all secondary rubber reclaimed is recycled and substituted in the primary rubber market directly.

Calculation of the effect of this increased level of secondary production on the energy requirements of the EUR 10 can be made on the basis that all phsyically available secondary resources are directed for recovery.

Home Production

| | | |
|---|---|---|
| $H_{np}$, | home primary natural rubber production | 0.70 Mt |
| $H_{ns}$, | home primary synthetic rubber production | 1.30 Mt |
| $H_{sp}$, | home secondary natural rubber production | 0.17 Mt |
| $H_{ss}$, | home secondary synthetic rubber production | 0.33 Mt |

Imports

| | | |
|---|---|---|
| $I_{nrp}$, | imports of natural rubber products | 0.18 Mt |
| $I_{nrr}$, | imports of natural rubber raw | 0.45 Mt |
| $I_{nrw}$, | imports of natural rubber waste | 0.01 Mt |
| $I_{srp}$, | imports of synthetic rubber products | 0.22 Mt |
| $I_{srr}$, | imports of syntehtic rubber raw | 0.52 Mt |
| $I_{srw}$, | imports of synthetic rubber waste | 0.02 Mt |

Further recovery potential (total physically available resources) have been identified at a level of 1.15 Mt total made up of 68% synthetic and 32% natural rubbers.

$$R_{pn} = 0.37 \text{ Mt}$$
$$R_{ps} = 0.78 \text{ Mt}$$

Assume relative energy values of the materials as:

| | |
|---|---|
| $E_{np}$, energy of primary natural production | 1.00 |
| $E_{sp}$, energy of primary synthetic production | 4.06 |
| $E_{ns}$, energy of secondary natural production | 0.34 |
| $E_{ss}$, energy of secondary synthetic production | 2.34 |
| $E_{sr}$, energy of synthetic rubber raw | 4.06 |
| $E_{nr}$, energy of natural rubber raw | 0.90 |

i.e. energy consumption of primary synthetic rubber production is 4.06 times as energy intensive as natural rubber due to the inclusion of petroleum feedstock in the product. Energy of secondary synthetic production is 2.34 times as energy intensive as primary natural production whereas secondary natural production leads to an energy saving of 66% over primary natural production.

The significance of the energy content of the imported materials can be calculated knowing the relative displacement of home and imported primary production and imported secondary production of both types of rubber, which is:

| | 1981 data (Mt) | Displaced by fraction of $R_p$ (Mt)* | Corrected production and import levels (Mt) | Denoted |
|---|---|---|---|---|
| $H_{np}$ | 0.70 | 0.19 | 0.51 | $H_{np2}$ |
| $H_{ns}$ | 1.30 | 0.49 | 0.81 | $H_{ns2}$ |
| $I_{nrp}$ | 0.18 | 0.05 | 0.13 | $I_{nrp2}$ |
| $I_{nrr}$ | 0.45 | 0.12 | 0.33 | $I_{nrr2}$ |
| $I_{nrw}$ | 0.01 | – | 0.01 | $I_{nrw2}$ |
| $I_{srp}$ | 0.22 | 0.08 | 0.14 | $I_{srp2}$ |
| $I_{srr}$ | 0.52 | 0.20 | 0.32 | $I_{srr2}$ |
| $I_{srw}$ | 0.02 | 0.01 | 0.01 | $I_{srw2}$ |
| Total | 3.40  1.34 N 2.06 S | 1.15  $0.37R_{pn}$ $0.78R_{ps}$ | 2.26 | |

*     Assumes that natural can only substitute for natural and synthetic for synthetic rubbers and that recovery reflects patterns of rubber consumption in EUR 10 in 1918. i.e. 68% synthetic, 32% natural.

The relative energy content of the imported materials can be calculated as:

$$\frac{(I_{nrp2} \times E_{np}) + (I_{nr2} \times E_{nr}) + (I_{nrw2} \times E_{nr}) \times (I_{srp2} \times E_{sp}) + (I_{srp2} \times E_{sp}) + (I_{srr2} \times E_{sr}) + (I_{srw2} \times E_{sr})}{(R_{pn} \times E_{ns}) + (R_{ps} \times E_{ss}) + (H_{np2} \times E_{np}) + (H_{ns2} \times E_{ns}) + (I_{nrp2} \times E_{np}) + (I_{nrr2} \times E_{nr}) + (I_{nrw2} \times E_{nr}) + (I_{srp2} \times E_{sp}) \times (I_{srr2} \times E_{sr}) + (I_{srw2} \times E_{sr})}$$

$$= \frac{(0.13) + (0.33 \times 0.90) + (0.01 \times 0.90) + (0.14 \times 4.06) + (0.32 \times 4.06) + (0.01 \times 4.06)}{(0.37 \times 0.34) + (0.78 \times 2.39) + (0.51) + (0.81 \times 4.06) + (0.13) + (0.33 \times 0.90) + (0.01 \times 0.90) + (.32 \times 4.06) + (0.01 \times 4.06)}$$

$$= \frac{2.34}{8.09} = 0.29$$

i.e. the potential energy savings which could be achieved by the recycling of an extra 0.37 Mt of natural rubber and 0.78 Mt of synthetic rubber intra EUR 10 must be reduced by 29%. This is the proportion of the energy involved in the provision of materials initially was expended extra EUR 10 in providing synthetic and natural:

o       raw products
o       rubber products
o       rubber scrap

Furthermore the increased level of secondary production has displaced proportionate levels of home primary materials production and imported materials.

235

LIST OF CONTACTS

The following abbreviations have been used for the Member States:

| | | |
|---|---|---|
| Belgium | – | B |
| Denmark | – | DK |
| Federal Republic of Germany | – | D |
| France | – | F |
| Greece | – | G |
| Ireland | – | IRL |
| Italy | – | I |
| Luxembourg | – | L |
| Netherlands | – | NL |
| United Kingdom | – | UK |

Information was also gathered from several other key countries in the field of wastes recycling and energy recovery:

| | | |
|---|---|---|
| Japan | – | J |
| Norway | – | N |
| Spain | – | E |
| Sweden | – | S |
| Switzerland | – | CH |

---

* Wherever possible, data for Spain has been included, but no specific enquiries have been made.

**F.13**     <u>Governmental or Institutional Contacts</u>

| | |
|---|---|
| OVAM, (Openbare Afvallstoffenmaatschappi voor het Vlaamse Gewest), Mechelen | B |
| Vrije Universiteit, Brussels | B |
| Centre Nationale de la Documentation Scientifique et Technique, Brussels | B |
| | |
| National Agency for Environmental Protection | DK |
| Technical University Lynngby, Copenhagen | DK |
| | |
| Umweltbundesamt, Berlin | D |
| Federal Economics Ministry, Bonn | D |
| Federal Statistics Unit, Wiesbaden | D |
| Federal Ministry of Interior, Bonn | D |
| TU — Hamburg Harburg | D |
| TU — Berlin | D |
| | |
| INSA, Villeurbanne | F |
| ANRED, Angers | F |
| OECD, Paris | F |
| CEPAC, Paris | F |
| Rylem Committee, Paris | F |
| Ministère de l'Environnnement | F |
| | |
| Paraskevopoulous & Georgiadis Ltd, Athens | G |
| | |
| Department of the Environment, Dublin | IRL |
| Institute for Industry, Research & Standards | IRL |
| National Institute for Physical Planning and Construction Research (An Foras Forbatha) | IRL |
| | |
| Ministry of Environmental Cultural Heritage, Rome | I |
| Ministry of Industry, Rome | I |
| Consiglio Nationale Delle Ricerche | I |
| | |
| Ministry of the Environment | L |
| | |
| TNO, Appledoorn-Zuid | NL |
| VROM, Department of Energy | NL |
| | |
| Building Research Establishment, Watford | UK |
| Brunel University, Uxbridge | UK |
| Department of the Environment, London | UK |
| Department of Industry, London & Manchester | UK |
| Department of Energy, London | UK |
| Harwell Laboratory, UKAEA, Didcot | UK |
| Energy Technology Support Unit, D. of Energy, Didcot | UK |
| Warren Springs Laboratory, Department of Industry | UK |
| CIPFA, London | UK |
| Offices of European Commission, London | UK |
| Open University, Milton Keynes | UK |
| University College of North Wales, Bangor | UK |
| | |
| Bureau of Mines, Washington | USA |

**F.2**    **Ferrous Metals Industry**

| | |
|---|---|
| British Scrap Federation | UK |
| British Steel Corporation, Croydon & Swansea | UK |
| Iron and Steel Statistics Bureau, Croydon | UK |
| Metal Bulletin, London | UK |
| Metals Society, London | UK |
| Metal Box PLC, Reading | UK |
| World Metal Statistics Bureau, London | UK |

**F.3**    **Non-Ferrous Metals Industry**

| | |
|---|---|
| Verband Metallverpackungen, Düsseldorf | D |
| Zinc Development Association, London | UK |
| Alcoa Ltd, Droitwich Spa | UK |
| Aluminium Federation, Birmingham | UK |
| British Aluminium Company, Gerrards Cross | UK |
| British Non-Ferrous Metals Association, Wantage | UK |
| British Secondary Metals Association | UK |
| Copper Development Association, St Albans | UK |
| International Lead & Zinc Studies Group, London | UK |
| International Wrought Copper Council | UK |
| Lead Development Association, London | UK |
| Metal Bulletin, London | UK |
| Metals Society, London | UK |
| Tin Research Institute, Greenford | UK |
| World Metal Bulletin | UK |
| World Metal Statistics Bureau, London | UK |
| Arthur D. Little, Cambridge, Mass. | USA |
| United States Bureau of Mines, Washington DC | USA |

**F.4**    **Paper and Board Industry**

| | |
|---|---|
| Verband Deutsche Papierfabriken, Bonn | D |
| CEPAC, Paris | F |
| OECD, Paris | F |
| Federatie Herwinning Grondstoffen | NL |
| British Waste Paper Association | UK |
| Industry Committee for Packaging & Environment | UK |
| Paper & Board Industrial Federation, London | UK |
| Paper Industries Research Association | UK |
| Metal Box PLC, Reading | UK |

F.5      **Wood and Timber Industry**

UN/FAO Timber Section, Geneva                                      CH

EBS - Holzcraft, Wackersdorf                                      D
Hans Kneuth Industrieanlage, Meersburg                           D

University College of North Wales, D. of Forestry, Bangor        UK
Association of British Plywood & Veneer Manufacturers            UK
Building Boards Federation                                        UK
Forestry Commission, Westonbirt Arboretum                        UK
Timber Research & Development Association, High Wycombe           UK
Caber Board & Co Ltd, Bannockburn                                UK
Sundeala Board Co Ltd, Sunbury                                   UK
Processed Woodchip & Woodflour Association, London               UK
Weyroc Ltd, Hexham                                               UK

F.6      **Glass Industry**

FEVE, European Glass Container Federation, Brussels              B

Studiengruppe Altglas der Bundesverband Glasindustrie,
    Düsseldorf                                                   D

Glass Manufacturers Federation, London                          UK
United Glass, Staines                                           UK

F.7      **Plastics Industry**

Association of Plastics Manufacturers Europe, Brussels          B

Verband Kunstofferzeugende Industrie e.V. Frankfurt-am-Main D

TNO-Plastics & Rubber Research Institute, Apeldoorn,           NL

Borg Warner, Leamington Spa                                    UK
British Plastics Federation, London                           UK
Metal Box PLC                                                 UK
Rubber & Plastics Research Association, Shrewsbury             UK
Rubber & Plastics Reclamation Association, Huntingdon          UK

F.8      **Rubber Industry**

Avon Rubber Co Ltd                                             UK
British Rubber Manufacturers Association, London               UK
Dunlop Ltd                                                    UK
International Rubber Studies Group                             UK
Rubber Association of Great Britain, London                    UK
Rubber & Plastics Research Association, Shrewsbury             UK
Rubber & Plastics Reclamation Association, Huntingdon          UK
Foster-Wheeler (Power Products) Ltd, London                    UK

**F.9    Chemicals Industry**

| | |
|---|---|
| Belgian Petroleum Federation, Brussels | B |
| Biolux, Mechelen | B |
| | |
| CEFIC, Paris | F |
| Centre Professionel des Lubrifiants, Paris | F |
| Syndicat National des Fabricants Raffineurs d'Huiles de Graissage, Paris | F |
| | |
| AMMRA (Waste Oils), Hamburg | D |
| | |
| Aschimia, Milan | I |
| Viscolube, Milan | I |
| | |
| British Chemicals & Dyestuffs Association, London | UK |
| British Petroleum, London | UK |
| Chemical Industries Association, London | UK |
| Chemical Recovery Association, Birmingham | UK |
| Frogson (Waste Oils) Ltd., Sheffield | UK |
| Imperial Chemicals Industry PLC, London | UK |
| Institute of Chemical Engineers | UK |
| Institute of Petroleum, London | UK |
| Kodak Ltd, Hemel Hampstead | UK |
| National Coal Board, Cheltenham | UK |
| Paint Manufacturers & Allied Trades Association | UK |
| Shell (UK) Ltd, London | UK |
| UK Petroleum Industry Association, London | UK |
| Waste Lubricating Oils Ltd., Stoke-on-Trent | UK |

**F.10    Secondary Materials & Commodities Recycling Associations or Federations**

| | |
|---|---|
| BESWA, Belgian Solid Waste Association, Brussels | B |
| BIR, Bureau International de la Recuperation, Brussels | B |
| International Solid Wastes Association, Brussels | B |
| Koniklijke Vlaamse Ingenieursvereniging, Antwerp | B |
| | |
| Aktion Saubere Landschaft e.V., Bonn | D |
| | |
| Federatie Herwinning Grondstoffen, Den Haag | NL |
| | |
| British Scrap Federation, Huntingdon | UK |
| Federation of Reclamation Industries | UK |
| INCPEN, London | UK |
| Maclaren Publishers Ltd (Materials Reclamation Weekly) | UK |
| National Federation of Demolition Contractors, Leicester | UK |
| National Industrial Materials Recovery Association | UK |
| | |
| National Association Recycling Industries, New York | USA |

**F.11**     <u>Thermal Conversion</u>

| | |
|---|---|
| Catholic University of Louvain (Professor H. Naveau, M. Demuynck) | B |
| Fläkt | B |
| Vrij Universiteit, Brussels (Professor A. Buekens, Ir. J. Willcox) | B |
| | |
| Babcock Kraus Maffei Industrieanlage GmbH | D |
| Berlin Technical University (Dr. Neumann) | D |
| Dornier Systems GmbH | D |
| Energieversorgung Schwaben AG | D |
| Federal Research Institute for Agriculture, Braunschweig | D |
| Hazemag GmbH | D |
| Umweltbudesamt | D |
| Hannover Municipal Authority (Dr. Knobloch) | D |
| Kiener Pyrolyse Gesellschaft mbH | D |
| Stadtwerke Rottweil | D |
| TU Hamburg-Harburg (Dr. Stegmann) | D |
| | |
| Bruun and Sorensen | DK |
| Kolding Kommunes | DK |
| | |
| ANRED | F |
| Andco Torrax/Calique | F |
| CPM/Europe S.A. | F |
| Carene S.A. | F |
| Compagnie Generale de Chauffe | F |
| Ecole Superieure des Mines de St.Etienne | F |
| France Dechets S.A. | F |
| Hydromer | F |
| INSA, Lyon | F |
| SOBEA | F |
| SIDUIC | F |
| Societe Francaise de Pyrolyse | F |
| Lyon I University | F |
| Valorga S.A. | F |
| Universite des Sciences et Techniques du Laguedoc | F |
| | |
| AMNU, Milan | I |
| BS Smogless, Milan | I |
| E.Bi.A, (Pavia) Broni | I |
| Industry and Ecology Commission, Ministry of Industry | I |
| Instituto di Chimica Fisica, Universita Milano | I |
| Ministry of Interior | I |
| Technical Polytechnic of Milan | I |
| SpA, Rome | I |
| Sogein | I |
| | |
| Department of the Environment | IRL |
| Institute for Industrial Research and Standardsd | IRL |
| National Board for Science and Technology | IRL |
| | |
| Ebara Infilco Company Ltd. Tokyo | J |
| Tokyo Metropolitan University (Professor Hirayama) | J |

243

| | |
|---|---|
| Fläkt | N |
| | |
| Esmil International BV | NL |
| TNO, Central Technical Institute, Apeldoorn | NL |
| | |
| AB Borlange Industriverk | S |
| AB Eksjö Energiverk | S |
| PLM Miljoteknik AB | S |
| SYSAV | S |
| RVF | S |
| Sockerbolaget (SSA) | S |
| | |
| Buhler Brothers Ltd. | CH |
| Ofag Zurich | CH |
| K & U Hofstetter AG | CH |
| | |
| Beverly Chemical Engineering Ltd. | UK |
| Blue Circle Group PLC | UK |
| Bootham North Ltd. | UK |
| Buhler Miag (England) Ltd. | UK |
| CLEAR (Cardiff Laboratories for Energy & Resouce Ltd.) | UK |
| California Pellet Mills | UK |
| Foster Wheeler Power Products Ltd. | UK |
| Henley Burroughs Ltd. | UK |
| London Brick Landfill Ltd. | UK |
| Motherwell Bridge Tacol Ltd. | UK |
| Peabody Holmes Ltd. | UK |
| Robert Jenkins Systems (CONSUMAT) | UK |
| Stone-Platt Fluidfire Ltd. | UK |
| Tollemache Ltd. | UK |
| Thomas Graveson Ltd. | UK |
| Waste Management Consultants Resource Recovery Ltd. | UK |

Printed and bound by CPI Group (UK) Ltd, Croydon, CR0 4YY

17/10/2024

01775690-0012